THE SECRET LIFE OF CATS

Out of the Wild and Into Our Homes

Carrie Arnold

NATIONAL GEOGRAPHIC

WASHINGTON, D.C.

Previous pages:
Sugar Pie, a seven-week-old gray-and-white domestic shorthair kitten, investigates her shadow in a sunny window bed.

Florida panthers, like this one at Tampa's Lowry Park Zoo, don't roar. They communicate through distinct sounds such as purrs, hisses, growls, and whistles.

CONTENTS

INTRODUCTION

A cat first adopted me when I was 23.

I had begged my parents for a cat as soon as I could talk, but severe allergies in the family meant that I had to wait until I moved out. That day finally came, and I began my search at local shelters. From the moment we met, Aria had me wrapped around her furry little paw. She was a tortoiseshell, or "tortie," tabby who spent much of our first six months together hiding under furniture. She had been a pregnant stray living behind a bowling alley before being rescued. Aria was an emotional basket case—and so was I. It was a perfect fit. With a lot of patience (and treats), I earned her trust. Soon, we were completely inseparable. What made my girl so endearing was that I knew her love was genuine. It was also selective, and she never feigned affection. She tolerated my parents (and, later, my husband), but she never pretended any of them were *her* human. And when she needed some time alone, she didn't hesitate to take it.

Aria was truly a cat's cat. Every endearing trait or annoying peccadillo that people normally associate with cats—aloof, charming, temperamental, fickle, loving—she had in spades. It was one of the reasons I loved her, and one reason I think humans love cats in general. They don't just give us their love. We earn it. They are pint-sized predators who curl up on our laps, casually dispatching mice and bugs into piles of gore, while simultaneously refusing to get their paws wet. They demand food at 4 a.m. and will also purr us to sleep when we're upset.

HUMANS LOVE CATS BECAUSE THEY DON'T JUST GIVE US THEIR LOVE. WE EARN IT.

No one was less shocked than I was to learn that cats, like Aria, had domesticated themselves. It's an epically cat move. It's not fair to say that cats are just solitary dogs with pointy ears. Scientists are just beginning to understand the rich complexity of what it means to be feline. Looking at cats through canine-tinted glasses makes them look antisocial and unintelligent. Cats have their own reasons for doing things.

By trying to look at Aria's world through her eyes and walk a mile in her paws, I realized that many of her inscrutable behaviors could be

Enjoying the tranquility, a cat rests near a vibrant garden in Stratford-upon-Avon, England.

comprehensible. Forcing myself to peer inside her secret life helped me enrich her days, just as she did mine. I made window perches so she could chitter *eckeckeck* at wildlife. Other cats were threatening, so Aria remained an only kitty. And I learned when I could pet her plush tummy and when it was just a Venus cat trap.

Cats will probably always remain a bit mysterious to us. Aria certainly has, even long after she crossed the rainbow bridge. Cats are likely glad to remain a bit secretive, but for those of us willing to dig deeper, we can share with them a wonderful world where both our species can lovingly thrive.

CHAPTER ONE

CATS THEN AND NOW

THEIR EVOLUTION AND BIOLOGY

A remote camera captures a cougar walking a game trail above the Buffalo Fork of the Snake River in Grand Teton National Park, Wyoming.

THE FIRST FELIDS

The story of cats begins long before the first kitty stalked a mouse or meowed for a head scratch. Biologists have traced most living mammals in the order Carnivora to a small group of mammals called miacids that lived 33 to 65 million years ago. These small ground- and tree-dwellers with long, lean bodies and short, stubby legs lived in the forests of North America and Eurasia. This order of flesh-eating mammals includes the family Canidae, with canids such as dogs, coyotes, wolves, and foxes, as well as the family Felidae, to which all felids, or cats, belong.

The oldest known felid, *Proailurus*, didn't appear in the fossil record until 30 million years ago and lived in what is now Europe. They had already evolved some of the hallmarks of their feline descendants. *Pseudaelurus*, a successor to *Proailurus*, appeared about 20 million years ago and was the first felid to venture into North America. We would be able to recognize *Pseudaelurus* as a true cat, with its slender body, short snout, rudder-like tail, and agile paws suited for climbing trees. These species included the house cat–size *Pseudaelurus transitorius*, the lynx-size *Pseudaelurus lorteti*, and the cougar-size *Pseudaelurus quadridentatus*.

The origins of living felids—descendants of *Pseudaelurus*—remained shrouded in mystery for many years, but a discovery high in the Himalaya in 2010 gave scientists new insight into some of the first modern big cats. *Panthera blytheae* lived somewhere between 4.1 to 5.95 million years ago in the Tibetan Plateau. Scientists believe it resembled modern snow leopards in body and lifestyle. Although *P. blytheae* isn't the direct ancestor of

***Proailurus*, the oldest** known felid, scratches a tree in this illustration. This extinct mammal lived during the Oligocene and Miocene periods. Its fossils have been found in Mongolia, Germany, and Spain.

COOL CATS WE LOST

No story of the last ice age is complete without *Smilodon*, the saber-toothed tiger. Despite its common name, this felid is not closely related to modern tigers. Some scientists believe it was part of a group of extinct felines called Machairodontinae or saber-toothed cats, famous for their fearsome upper canines. But these oversize teeth may have been just for show. They lacked a protective covering of enamel, which means *Smilodon*'s teeth would have broken easily.

Recent research offers evidence that *Smilodon* was a forest dweller who feasted on smaller herbivores that ate leaves, ambushing them in closed forests rather than chasing them in more open habitats. This suggests that *Smilodon* had limited prey options, which could have sealed its fate when its environment changed. The exact reason why *Smilodon* died out is unknown, but theories include extinction of its prey and climate change.

most living felids, the discovery of the species may confirm the Asian origin of modern species of the genus *Panthera*, but not of all modern cats.

Cats Today

Biologists divide all modern cats of Felidae into two subfamilies, Pantherinae and Felinae, and 14 genera. Native felids live almost everywhere on Earth except Australasia, Antarctica, and the Arctic, stalking their prey in forests, savannas, deserts, and rocky outcroppings. Although lions, tigers, and leopards might be the most famous of the living felids, scientists have identified more than 40 species and 77 subspecies of cat alive today.

While there are many similarities among felids, cats come in all shapes and sizes, and each species has its own unique characteristics. For example, the pint-sized, nocturnal sand cat lives in the Sahara, the Middle East, and parts of Central Asia, camouflaging itself in its desert home with pale sandy to light brown-gray fur, broad face, and large ears. There's the medium-size caracal, who uses its long, powerful legs to leap up to 10 feet (3 m) in the air to snag a bird on the wing, though it will eat any prey it can catch, including rodents, monkeys, and impalas. Across Central Asia, the stocky, shaggy Pallas's cat survives on pikas and other rodents, using its dense fur as protection from the cold. And the forests of Central and South America are home to the margay, a medium-size, lean felid with a spotted coat that hunts at night and sleeps during the day.

The sand cat, *Felis margarita harrisoni* (below), is native to North Africa, the Arabian Peninsula, the Middle East, and Central Asia.

CAT CONSERVATION CRISIS

AS A RESULT of poaching and a loss of habitat and prey, populations of many large wildcats are plummeting. Although Amur leopard (right) numbers appear to be increasing, only 50 of these large spotted beauties remain in the Russian Far East. They are threatened by poaching for their striking spotted fur, habitat degradation, and other factors. Around the world, other cat species are also at risk as populations shrink and fragment, leaving the remaining animals vulnerable to inbreeding and disease.

But conservation efforts offer hope. About a thousand Iberian lynxes (above) now live in Spain and Portugal (up tenfold in 20 years). Their resurgence was fostered by reintroducing captive-bred lynx, rewilding lynx ranges, boosting prey species, and creating wildlife corridors and highway tunnels.

One of the most important ways to protect these at-risk felines is by saving their prey. If the hunting of their prey is allowed, it needs to be done sustainably. These strategies will help humans and cats thrive together.

CHARACTERISTICS OF CATS

Modern cat species can range in size from the massive Siberian tiger, *Panthera tigris altaica,* which can weigh more than 700 pounds (317 kg), to the dainty and diminutive black-footed cat, *Felis nigripes,* of southern Africa, and the rusty-spotted cat, *Prionailurus rubiginosus,* of India and Sri Lanka—each of which weighs in at about 2 to 4 pounds (0.9 to 1.8 kg). Despite more than a hundredfold difference in size, all cats share certain physical traits. They're what make a cat a cat. A 2013 study showed that the Siberian tiger and the domestic cat share 95.6 percent of their DNA. This similarity may be reflected in these cats' common need to find and capture prey.

IF I FITS, I SITS

One thing nearly all domestic cats share is a love of boxes. Humans have photographed cats sitting in all kinds of containers, even those far too small for a full-grown kitty, tagging them with the meme title "If I fits, I sits."

A new study of illusory contour susceptibility shows that cats love these spaces so much, they'll even sit in ones created by visual illusions. Whether the boxes were real or illusory, the cats overwhelmingly preferred sitting in boxlike shapes on the floor. As for why cats love boxes, well, that's less clear. Scientists hypothesize that boxlike enclosures help kitty feel safe and secure. Cats in boxes can hide from real or imagined predators who can't sneak up on them. As ambush predators themselves, they can stalk prey (real or imagined) from boxes. And for creatures whose instinct when stressed or fearful is often to run and hide, boxes seem ideal.

DESPITE DIFFERENCES IN THEIR SIZES, ALL CATS SHARE CERTAIN PHYSICAL TRAITS.

Predator Senses and Physique

Felids have acute sight aided by their large, forward-facing eyes, which allow them binocular or stereoscopic vision to evaluate distances, as well as highly dilatable pupils for excellent night vision. Their senses of smell and hearing are also acute, enabling them to identify territories and find mates, as well as locate their next meal. Their physique is specialized to their predator status as well. Felids have strong but flexible spines and specialized musculature that gives them high speed, agility, and strength. Strong hind legs let felines jump to great heights and rapidly accelerate in pursuit of prey.

Siberian or Amur tigers, *Panthera tigris altaica,* are an endangered species native to the Russian Far East and parts of northeastern China and North Korea.

Cats have only 28 to 30 teeth, compared with 48 to 50 in dogs, giving them the fewest teeth among all carnivores. Felids use fang-like upper teeth and equally sharp ones on their lower jaws to rip flesh from bone, swallowing chunks of meat whole. Their short, muscular jaws give cats an intensely powerful bite.

Bodies at Work

All cats share a quiet, stealthy gait, thanks to the fact that they walk on their toes, which are protected by soft pads. And although all felids have claws to help them climb trees, defend themselves, and subdue prey, they typically have an elastic ligament in each toe to keep their claws retracted. Because they don't rub against the ground, a cat's claws are always sharp and ready for action.

A cat's body has evolved to help it stalk and then chase down that next meal. Rapid contraction cells in their muscles let felids quickly reach top speed. This type of cell promotes power and acceleration, but only in short bursts, making them more like a sprinter than a marathon runner.

A favorite resting spot for lion cubs are termite mounds, like this one in the Okavango Delta of Botswana (right).

Stalking prey, an orange tabby (opposite) hunts among the wild-flowers near the Philosopher's Path, Kyoto, Japan.

In contrast, slow-twitch muscle cells help felids slink through grass and lie in wait for long periods while stalking prey.

Cats' digestive systems have an abbreviated small intestine that allows them to maintain a sleek profile so that they can move quickly and agilely when pursuing prey. They also have enzymes that let them digest uric acid in meat.

Most wildcats—with a few exceptions, including the bobcat, lynx, and caracal—have tails at least one-third to one-half their body length. Besides helping keep them warm on a cold night, tails can act as rudders and balancing aids as cats move through trees and rugged terrain, making quick turns to chase prey.

Many cats are crepuscular, meaning they are most active at dawn and dusk, when their prey is also out and about. When not actively hunting, domestic cats often sleep for up to 13 to 16 hours each day. This helps them conserve energy for their next chase. Cats can awaken in seconds to allow them to pursue an animal that has happened by.

And what would a cat be without its rumbling purr or ferocious roar? Scientists used to think cats that purred couldn't roar, and vice versa, because of differences in a specialized bone in the throat called a hyoid. But newer research shows that the roar/purr divide is actually due to differences in the structure of the cat's larynx rather than the hyoid.

MANY CATS ARE MOST ACTIVE AT DAWN AND DUSK, JUST LIKE THEIR PREY.

With only tiny variations, this intricately adapted cat body allows felids to live in forests and deserts, savannas and tundras, in the trees, and along riverbanks. And it's also shared by the cats who have found their way into our homes and hearts.

OUT OF THE WILD AND INTO OUR HOMES

Anyone who has shared a home with cats knows they make great companions. But unlike dogs, cows, horses, pigs, and other animals, cats weren't the obvious choice for domestication. Wildcats weren't large enough to transport humans and carry heavy loads, nor were they especially valuable for their milk and meat. Cats weren't as easily trainable as dogs and were less inclined to follow commands. Cats also weren't herd or pack animals that lived in groups, where a human could replace an alpha animal.

IN A MOVE THAT'S JUST SO LIKE THEM, CATS DOMESTICATED THEMSELVES.

Cats are also obligate carnivores, which means they have a limited ability to digest anything other than meat. Living in close quarters with an animal that might consider you dinner isn't generally a recipe for success. A cat's carnivorous stream presents other problems, too. Their food preferences limit their diet, because they can't get the nutrients they need from other sources. It also places them in direct competition with early humans for these meaty morsels. In short, cats seemed far from the ideal choice for pets for humans—and, in fact, they weren't. Instead, in a move that is perfectly on-brand for cats, our feline friends actually domesticated themselves.

In Search of Food

When humans first began farming thousands of years ago, they found themselves with a steady surplus of grain for the first time. The extra food brought security to the new farmers, but it also brought pests that weren't about to walk away from such a starchy smorgasbord. Mice and rats feasted on stored grain, not only chowing down on necessary food stores but also spreading deadly disease in their wake. This created a conundrum for the new agriculturalists: how to keep the grain but eliminate the pests.

Though humans may have lived alongside wildcats, the two species probably had limited interactions. Cats that don't interact with humans as kittens remain skittish of the clumsy bipeds who beckon "Here, kitty, kitty." What's more, cats are most active at dawn and dusk rather than during the day, like humans. Even in places like eastern

Some feline family members share the shade with a cat lover on Symi Island, Greece (opposite). Socializing cats and kittens through play, treats, petting, and cuddling is fun for them and for us.

For cats at any age, hunting a live or catnip mouse never gets old (below).

and southern Africa, humans may have lived alongside wildcats, such as the African wildcat *(Felis silvestris lybica)* but had limited contact with these felines. But agriculture brought wildcats ever closer to humans, because the new farming endeavor promised cats brave enough to venture near human settlements the possibility of an all-you-can-eat rodent buffet.

These furry pioneers found themselves dining not only on a mouse bonanza but also food scraps left out by humans, whether accidentally or deliberately. Human settlements may have also offered some protection from predators, making it more likely that the offspring of these friendly felines would survive to adulthood. Humans who learned how to live with their new furry

Some domestic cats are more social than others. And like humans, sometimes even friendly ones just need to chill (right).

A barn kitty walks the perimeter on his Nebraska ranch (below). Working cats control pests and are often boon companions for other animals.

friends may have had more grain, thanks to fewer rodents gorging on their harvest. Cats also killed other pests that posed continual threats, such as snakes. Cats were rewarded with food and shelter from humans, and we gained excellent, low-maintenance pest control that purred.

Modern domesticated cats may be considered less social than dogs and even livestock like sheep, pigs, and goats, but that doesn't mean they haven't adapted to life near humans. In comparison to wild felids, such as snow leopards, some house cats are significantly more social and seek out other cats, animals, and humans, while others are semisolitary. Their overall behavior, scientists found, varies but tends to be more social than that of the solitary hunters from which they descended. Cats have adapted. Being able to tolerate living in close quarters with humans and even other cats may have given early domestic cats a major edge.

Cats Come Home

Human settlements that built up around agricultural centers may have played a role in helping cats start the process of domestication. When scientists added hiding spaces to the cages of captive leopards, they found the cats were less stressed and explored more. So the multitude of nooks and crannies in human-built environments, such as barns, silos, and other outbuildings, may have helped early domesticated cats cope with the difficulties of living in close quarters and offered a retreat when the outside world grew too challenging. Without litter boxes, however, cats would have required free access to the outdoors to do their business, as well as find mates. This allowed cats to keep one paw in their wild past while taking a tentative step toward a more domesticated future.

Although the benefits of domestication for both cats and humans may have been obvious, the details of exactly which wildcat species were tamed, as well as when and where this happened, once remained a mystery.

Origins of Domestication

Genetic evidence points to the Fertile Crescent in the Middle East as being the cradle of cat domestication. When geneticists compared the DNA of domestic house cats with that of various subspecies of *Felis silvestris* in Europe, Africa, and Asia, they found that house cat genomes were nearly indistinguishable from *Felis silvestris lybica* in Africa and the Middle East, but clearly separate from *F. s. silvestris* in Europe,

F. s. bieti in China, *F. s. ornata* in Central Asia, and *F. s. cafra* in southern Africa, as well as any other subspecies of wildcat. This indicates that house cats arose in the Middle East and not in other places where humans may have lived in close contact with wildcats.

In 2017, a more detailed analysis of feline DNA from the last 9,000 years—including everything from ancient Romanian cats to modern African wildcats—revealed two main genetic lineages from *F. s. lybica* in modern domestic cats. The first came from the initial process of domestication in the Middle East. When agriculture emerged from the Fertile Crescent, Europe, and northern Africa, rodents followed, which in turn brought the cats. The second lineage came from the African cats that became popular in ancient Egypt. They were believed to bring their humans (royal or not) good luck while protecting their households from pests. In this way, they became working pets.

As humans traveled at sea and overland, rats and mice came with them. Whether cats followed for the access to rodents or for human companionship isn't clear, but scientists do know that cats accompanied humans as

The Lion Gate shielded the neo-Hittite temple of Ishtar at Ain Dara, Syria, for 3,000 years. It was severely damaged in January 2018 in an air strike.

Cats, like this one lounging near the Great Sphinx of Giza, have been fixtures in Egypt for thousands of years.

they migrated around the world. Cats descended from *F. s. lybica* ultimately outcompeted local wildcats. This means that almost anywhere you go in the world, the pet cats you find are descendants of a wildcat from the Middle East.

It wasn't their natural fitness and survival instincts that gave these cats an edge on domestication, but their skill in making their lives with humans work for them. Evidence from burials shows that humans provided for their furry friends as they developed dental disease or after they fractured bones. And in this way, with humans at their side, cats conquered the world.

Keeping It (a Little) Wild

Unlike other domesticated animals, however, cats continued to largely direct their own lives. Although cats benefited from human contact, they didn't depend on humans for survival. They also continued to choose their own mates, coming and going as they pleased. Their reasons for staying near us were distinctly their own and not something humans controlled. In many ways, this still is true. This means that, at best, cats are only partially domesticated. As a result, cats retain the wild body plan.

Wildcats have vertical stripes running down their bodies, perpendicular to the spine in a pattern called mackerel. Although this pattern still appears in modern house cats, the "classic" or blotched tabby pattern (with bold swirls along the cat's body) appears only in domestic cats. DNA evidence revealed that this coat pattern didn't appear until the Middle Ages.

CAT MATURATION: KITTY GROWS UP

Although their cuteness is timeless, a kitten becomes a cat in a vanishingly short period. Like many newborn wild felids, kittens are born blind, deaf, and almost completely helpless. They depend on their mother for food and shelter, and even to urinate and defecate. In less than a month, kittens learn to see, hear, and walk, all fueled by their mother's milk and care.

From Kittens to Young Adults

A kitten's first teeth, including canines and incisors, begin to appear at around three weeks of age, and more continue to arrive in its fourth and fifth weeks. Besides eating with them, cats also use their teeth to remove knots from their fur. Kittens will begin imitating their mother by licking themselves to keep clean.

Fully weaning off mother's milk can take several weeks. Some mothers will wean their kittens earlier, while others may let kittens suckle even after they stop producing milk. This comfort nursing can also occur on soft toys and blankets and is a way for cats to self-soothe (not unlike thumb sucking).

Sexual Maturity and Spaying/Neutering

On average, female kittens reach sexual maturity—when they go into heat and can become pregnant—at six months old. Male kittens reach sexual maturity at the same age or older, but when they do, they are generally larger than a newly sexually mature female kitten. This early age at reproduction means cats can breed, well, kind of like rabbits. During her reproductive life, one female cat can have more than 100 kittens.

Not all of these kittens will survive to adulthood. One 2004 study found that three-quarters of kittens born in the wild died or disappeared before they reached six months of age. Despite this, shelters are often overcrowded with cats and kittens. Every year in the United States, thousands of dogs and cats are euthanized by animal control services due to lack of shelter space, homes, and fosters for their care.

People who haven't lived with kittens sometimes ask: Are they always so crazy? Nope. They're just practicing to be grown-up cats. That includes jumping, biting, exploring, and hunting all things that move.

A queen (mother cat) can give birth to as many as a dozen or more kittens in her litter. The average number is four to six. First-time moms tend to have fewer kittens.

Many vets suggest waiting to spay or neuter until kitty is five or six months old, although some rescue groups will do this earlier as part of informed adoption initiatives. Once the kittens reach 2.2 pounds (1 kg) and eight weeks old, they are spayed or neutered and adopted into loving homes. Spaying and neutering not only keeps kittens from reproducing but also prevents female cats from going into heat, lowering the risks of mammary cancer later. It also extends their life span by decreasing a cat's urge to roam during a hormonal surge, which keeps them from fighting with other animals, getting hit by vehicles, and other behavioral mishaps both indoors and out.

A KITTEN BECOMES A CAT IN A VANISHINGLY SHORT TIME.

A CAT'S WORLD

To walk in a cat's paws, we first need to understand how a cat experiences the world. Although cats use the same senses as humans—sight, sound, smell, taste, and touch—they understand and process their inputs quite differently. The world according to cats, then, is very different from the one we travel through. But it also has some tantalizing similarities that can help us live more harmoniously with our furry friends.

Sight

Like humans, cats use their vision to see the world around them—and to hunt down their next snack. But differences between human and cat eyes mean that we see the world quite differently.

Although a cat's precision pouncing in the dark may make it seem that she has built-in night vision goggles, cats do need some light. But though a human's night vision is iffy at best, the dark is a cat's time to shine. Millions of years of evolution have made it more likely that many cats will be active and hunt at dusk and dawn.

Light enters the eye through the cornea, the round, transparent surface of the cat's eye. The cornea focuses light on the retina, which lines the back of the eye on the inside. A cat's cornea is large and dome-shaped, letting cat eyes gather

THIS IS YOUR CAT ON CATNIP

A single whiff of catnip might make a kitty bliss out or act wild—or it might have no effect at all. Around 50 to 75 percent of cats have the genetic ability to respond to *Nepeta cataria* (right), a leafy herb in the mint family. Researchers believe that nepetalactone, the cat-stimulating compound in catnip's essential oil, targets receptors in the feline brain that make cats happy. Cats who love to sniff catnip may go into paroxysms of rubbing, rolling, and flipping about. Other cats who eat it become mellow. (Though catnip is safe, vets suggest having kitty enjoy it in small amounts.) Cat behavior is delightfully altered by catnip's influence for about 10 minutes at a time.

Humans have benefited from catnip, too. As far back as the 1600s, people brewed it as a soothing tea, and they have also used it as a mosquito repellent because of its essential oil.

This domestic kitten *(Felis catus)* has one blue eye and one green eye, a genetic condition known as heterochromia that does not affect her vision.

the maximum number of photons—a key adaptation to their low-light living. Cat pupils are long and vertical, narrowing to a slit in broad daylight but expanding up to 300-fold when it's darker (human pupils only grow 15 times bigger). The backs of a cat's eyes have a layer called the tapetum lucidum, which reflects unabsorbed light back into the retinas, an adaptation to help the cat see in dim light, and causes eye shine, the glow that can be seen when light shines on them in the dark. Their peripheral vision is also better than ours.

Other aspects of feline vision are not as sharp. Because their retinas have fewer cones—the photoreceptors that perceive color—cats are thought to see the world less vibrantly and in fewer hues than humans. These cones are also responsible for sharpness of vision, so a cat's eyesight is blurrier, despite their superior low-light vision. What cats can see from 20 feet (6 m) away, we can see at 100 feet (30 m). But this doesn't slow them down. Cats respond more readily to motion than to the intricate details and colors of an image, so they aren't hindered by their reduced color vision.

CATS CAN PINPOINT THE LOCATION OF A SOUND TO WITHIN SEVERAL INCHES.

This subadult male caracal (opposite) was hand-raised by farmers, and he loved people. At Anysberg Nature Reserve in South Africa, he learned to live in the wild.

Sound

A cat's triangle-shaped ears act like small, furry satellite dishes. Their ear flaps, or pinnae, can independently rotate forward, backward, and sideways to zero in on a sound's location. The pinnae's 180 degrees of rotation means that cats can pinpoint the location of a sound to within several inches in just six-hundredths of a second—faster than the blink of an eye—from up to three feet (1 m) away. They can also distinguish extremely subtle differences in sounds, even as little as one-tenth of a tone.

But their ultrasonic hearing (far superior to humans and even dogs) doesn't mean that Beyoncé and Beethoven are to a cat's musical tastes. In 2015, a research team from two U.S. universities tested tunes incorporating feline-centric sounds that included purring and a pulse reminiscent of suckling. The results showed that cats preferred the cat songs ("Cozmo's Air" and "Rusty's Ballad") to music composed for people.

Smell

Smell, unlike the other four senses, is fully developed straight out of the cat womb. The newborn kitten quickly uses its nose to navigate toward the nearest

NOT JUST FOR TASTING

Cats love to groom themselves almost as much as they love to sleep, typically spending 30 to 50 percent of their waking hours cleaning their fur. The secret to their self-cleaning success? Curved, hollow-tipped spines on their tongues called papillae, which transfer large amounts of saliva from mouth to fur. This not only cleanses kitty right down to her skin but also lowers her body temperature as the saliva evaporates.

Researchers found that these scoop-shaped papillae inspire new approaches to cleaning and depositing fluids on hairy, furry, and fuzzy surfaces. David Hu of Georgia Tech compared a cat's tongue to "a very smart comb." His team developed the TIGR (Tongue-Inspired GRooming) brush, studded with small, curved flexible spines. The TIGR required less force to use compared to a human hairbrush and could be cleaned with just a finger swipe. It could even be used to make cats less allergenic.

HOW CATS ALMOST ALWAYS LAND RIGHT SIDE UP

CATS (ALMOST) ALWAYS land on their feet. This sounds like an urban legend. In *Falling Felines and Fundamental Physics*, Greg Gbur of the University of North Carolina at Charlotte outlines four hypotheses about how cats right themselves: the "tuck-and-turn," the "falling figure skater," the "bend-and-twist," and the "propeller tail." Gbur thinks cats use a combination of all these methods to land safely.

It's all the result of a vestibular system. This not only gives vertebrates a precisely tuned sense of balance and movement but also lets felines right themselves if they do slip and fall.

In the split second after kitty realizes *oops*, he uses cues about gravity detected by the inner ear to rotate his head to face right side up. Then, kitty will arch and twist his spine, extending his legs and pulling them in at different points (and directions) in his rotation. This places kitty in the ideal landing position.

1 **Flexion at the waist; front of body begins to turn**

2 **Tuck front legs, extend rear legs; front rotates ≤90°, rear rotates ≥10°**

3 **Extend front legs, tuck rear legs**

4 **Rotate rear more, rotate front less; repeat steps 2–4 for 180° rotation**

nipple and get its first sip of nourishing colostrum and milk.

Experts believe that a cat's sense of smell is about 14 times better than ours. A domestic cat's olfactory epithelium—the specialized tissue in the nose containing the receptors that detect odors—is five to 10 times larger than a human's. As a result, cats have up to 200 million specialized cells that detect smells, compared with our mere five million.

Our feline friends have another tool at their disposal, the Jacobson's organ. Tucked above the mouth, receptor cells in the Jacobson's organ connect to the part of the brain associated with sexual, feeding, and social behaviors. If cats smell something interesting, they will partially open their mouths and curl their upper lips, an expression called the Flehmen response. This shunts air molecules to the Jacobson's organ. The inhaled air is trapped on the olfactory epithelium and/or the Jacobson's organ, giving kitties an extra chance to detect scent molecules.

Touch

Whiskers on kittens may famously be one of our favorite things, but cats really depend on them. Known formally as a vibrissa, a whisker is longer and thicker than normal cat hair.

Each of a feline's whiskers grows from a follicle packed with nerves and blood vessels, making them as sensitive as human fingertips. These vibrissae help to compensate for a cat's less-than-stellar close vision. They detect subtle air movements that can indicate the presence of prey and help cats navigate around obstacles. ⦿

Cat whiskers are proportional to a kitty's body width (including fur), which helps in navigating tight spaces.

THERE'S NO FANCIER FEAST

If there's one thing cats love to eat, it's meat. Their meat-eating lifestyle isn't just a preference, it's a requirement. Cats are hypercarnivores, meaning that meat makes up more than 70 percent of their diet. Domestic cats whose food preferences were studied in a lab chose to eat large amounts of protein (a target intake of about 52 percent of their diet), moderate amounts of fat (36 percent), and minimal carbohydrates (12 percent).

A CAT'S HIGH PROTEIN NEEDS ARE EVIDENT FROM BIRTH AND INCREASE WITH AGE.

During their evolution, cats lost one of the genes responsible for the sweet taste receptor, which may have influenced their path to becoming hypercarnivores. Because cats don't eat sweets or starches, they did just fine without these receptors. This is not unique to cats. Three species of vampire bats, which exist on a diet exclusively of blood, have also acquired mutations in their sweet taste receptor that make it useless.

Cats have several adaptations that reflect their low-carb lifestyle. As a result of evolution, their salivary glands don't produce amylase, the enzyme that starts the digestion of carbohydrates. Overall, cats have lower amounts of carbohydrate-digesting enzymes throughout their digestive system.

CATS DON'T JUST HUNT FOR FOOD

Anyone whose family includes a cat can't ignore that their four-legged friend is also a furry predator. This hunting instinct doesn't disappear when a cat's physical appetite is sated. Cats are essential predators: They hunt and kill prey, even if they're not looking for food.

Researchers have found that even well-fed cats seek out prey regularly. Jekyll Island off the coast of Georgia is home to colonies of free-ranging cats provided with food and veterinary care by volunteers. But the island is also an important stopover for migrating birds. Scientists fixed small, collar-mounted video cameras on 31 cats and recorded life from their point of view for more than a year.

Hours of video footage revealed that 83 percent of the colony cats showed hunting behavior. More than 60 percent successfully killed prey, eating most of the animals they caught.

To protect wildlife, experts recommend keeping cats indoors and offering them ample play-hunting time.

Some protection in numbers: A lioness retreats before a large group of Cape buffalo in the Okavango Delta of Botswana.

A cat's high protein needs are evident from birth. Kittens need 1.5 times more protein per pound of body weight than puppies and other young animals. Adult cats need even more—two to three times more protein than adult omnivores. Nor can cats be flexible about their need for protein, which they use as a source of energy, as a building block for their own muscle, and as a source of essential amino acids (those protein building blocks that can't be made by the body and must come from food). If cats don't get enough protein in the food they eat, they can't reduce their body's requirement for this macronutrient. This makes them vulnerable to protein malnutrition.

Protein can be found in plants, but cats depend on animal protein. Only animal flesh contains several essential amino acids upon which cats have come to depend. Compared with omnivores, cats need higher amounts of the amino acids taurine, methionine, cysteine, and arginine. Without adequate amounts of these essential amino and fatty acids, cats can develop severe medical problems.

Taurine and arginine are found only in animal tissue. Methionine and

cysteine are found in plants, but not at high enough levels for feline diets. This means that cats can't meet their nutritional needs from vegetarian or vegan diets. Cats also need certain essential fatty acids, including arachidonic acid, which is needed for proper blood clotting. As omnivores, dogs have maintained their ability to synthesize these molecules from other components. It's why dogs can eat cat food and be healthy, but cats can't thrive on commercial dog food.

In the Dish

When it comes to what type of animal protein they eat, domestic cats can be relatively flexible. Their wild counterparts eat birds and small rodents like mice, shrews, and rats. But they also eat insects and reptiles, including spiders, grasshoppers, lizards, and snakes.

Free-ranging cats living in and around human settlements might meet their nutritional needs with people food. Well-meaning humans may actively feed feral cats living near their homes

Domestic cats can eat many kinds of animal protein. Knowing that the urge to hunt overcomes even the youngest kittens means we must keep an eye on them.

and workplaces. Even without food specifically for them, however, outdoor and community cats will scavenge from garbage containers and eat scraps left by humans. One study of outdoor cats in two areas of Brooklyn found that the neighborhood with more poorly contained trash and more abandoned buildings supported more cats than another neighborhood with better contained garbage and few abandoned buildings.

VETS AGREE: NON-MEAT DIETS AREN'T ENOUGH FOR KITTY'S NUTRITIONAL NEEDS.

Feral cats can dine quite happily on a wide range of foods, which may come as a surprise to those who live with their more finicky domesticated counterparts. A preference for the familiar and a more adventurous palate can each have their benefits and risks. Staying with the tried-and-true means less risk of eating something poisonous. Venturing out and trying something new, however, can let a cat take advantage of foods their companions won't touch.

Under ideal conditions, domestic cats will eat numerous small meals throughout the day, like their wildcat relatives. Whether wet food or dry, cats decide what to eat (and how much) based on the smell, texture, taste, and temperature of their food. Experiments have shown cats are especially drawn to foods with a high umami or savory flavor, indicating that the food contains protein. All of this together means that cats generally eat the right amount to stay healthy and happy, even when given free access to unlimited amounts of food.

Cats also need to find adequate water to drink. Like many animals with a history of desert living, cats can produce highly concentrated urine to conserve body fluids if drinking water is scarce. And they can tolerate large, short-term fluid losses of up to one-fifth their body weight comparatively well. But their thirst response is weaker than in dogs or humans and can lead to chronic dehydration as well as urinary tract and kidney disease.

Both domestic cats and wild ones like this bobcat tolerate thirst and fluid losses well, but not for long periods. Drinking enough fresh water daily is essential for their health.

CAT VARIETIES

Ranging from inky black to tawny gold and everything in between, a cat's fur coat helps it hide from animals it stalks and from its own predators. And cat coats don't just vary in color, they also vary in pattern. What would a tiger be without its stripes, or a leopard without its rosettes?

Cats have a broad range of fur colors, patterns, lengths, and textures. The variation is created by several genes. By mixing these genes in different combinations, felids can show a wide assortment of fur coloration traits. These physical characteristics, however, do not indicate a cat's breed. Even coat colors that are seemingly characteristic of one breed (such as the Siamese's colorpoint coat, with a light body and darker extremities) can occur in domestic cats of no particular pedigree.

Domestic cats have inherited this stunning array of genetically determined colors and patterns. The two primary coat color pigments are *eumelanin*, which produces black and brown, and *pheomelanin*, which makes red and yellow (or orange).

The genes that control fur color are located on the cat's X chromosome. Like other mammals, a cat with two X chromosomes (XX) is female, and a cat with an X and a Y chromosome (XY) is male. Male cats always get their coat color from their mother, whereas females get theirs from mom and dad. So male cats can be either orange or black, while females can be orange, black, or orange *and* black.

Female kitties from this latter group are called tortoiseshell cats (or torties), with their fur's orange and black colors randomly dispersed in a brindled

Ragdoll cats have sweet natures and their own distinctive beauty: colorpoint coats; blue eyes; sturdy, muscular bodies; and silky-soft longish fur.

A BIG DOSE OF TORTIE-TUDE

Besides their telltale coloring and good looks, many tortie lovers say their cats also have a big dose of attitude. It's why some consider them to be the divas of the cat world. Although temperament varies widely among cats, scientists from the University of California, Davis, surveyed more than 1,200 people who said their calicos and torties (and black-and-white tuxedo cats) were more likely to be aggressive during everyday interactions, handling, or vet visits than other cats. The most docile cats, as reported by their humans, were solid black, gray, and white. Researchers aren't sure why and don't believe behavior differences are caused by fur color itself. Instead, "cattitude" frequently appears in tricolor cats, just as redheads often have freckles.

Despite their generally placid behavior, solid black cats at shelters often wait the longest to be adopted, likely stemming from the stereotype that black cats are unlucky.

pattern, sometimes with cream or gold highlights. Because this color combination needs two X chromosomes, torties are nearly all female. Only about one in 3,000 torties are male, and this pattern only appears due to chromosomal abnormalities. A calico cat is similar to a tortie, but calicos also have a genetic condition called piebalding that produces patches of white fur. Their colors aren't brindled but are discrete patches of orange, black, and white.

Genes and Color Patterning

The calico patterning is set early in a cat's embryonic development. To prevent females from getting a double dose of genes on their X chromosomes, one copy is switched off, creating what's called a Barr body. If the X chromosome with the orange pigment gene is turned off, she will have a patch of black or brown fur. If the X chromosome with the black pigment gene is turned off, then orange pigment will show. As she matures, these patches enlarge, giving the calico her distinctive markings.

Other genes control the intensity of a kitty's fur coloring. The dominant dense pigment gene produces deep, vibrant colors. The recessive dilute gene lightens orange to cream and black to gray. The agouti gene determines the deposition of color in the hair. A single hair from a cat with the dominant form of this gene often has

Maine coons are known as "gentle giants." They are one of the oldest natural cat breeds in North America, and their fur can be a wide range of colors.

THE DOMINANT DENSE PIGMENT GENE PRODUCES DEEP, VIBRANT COLORS.

color bands. The agouti gene switches between depositing the two pigments in the hair shaft as it grows.

Tabby striping can appear with other color patterns. Tabby patterning can occur in brown, orange, and black/gray fur, as well as on the tortie (called torbie, for tortie-tabby), and the calico (caliby, for calico-tabby). Tabbies have dark, pencil-thin eyeliner fur and a distinctive dark, striped "M" on their foreheads.

Tabby pattern variation is the result of a gene called transmembrane aminopeptidase Q (dubbed *Taqpep*), which also causes a pattern of stripes and blotchy spots on the coat of a rare cheetah breed.

The ancestral tabby patterning is known as mackerel, with narrow stripes down the cat's flanks. The classic tabby has marbled or blotched stripes, and a butterfly pattern often appears on the shoulders. Despite the "classic" name, researchers found that this type of tabby marking didn't appear before the Ottoman Empire in the Middle Ages.

Genes also control whether a cat has short or long fur. The length gene governs how long a cat's fur is. The dominant form, *L*, codes for short fur, and the recessive *l* codes for long fur.

All orange cats, like these feral kittens, are tabbies. Their fur's stripes can be hard to see sometimes, but they're there. About 80 percent of orange kitties are male.

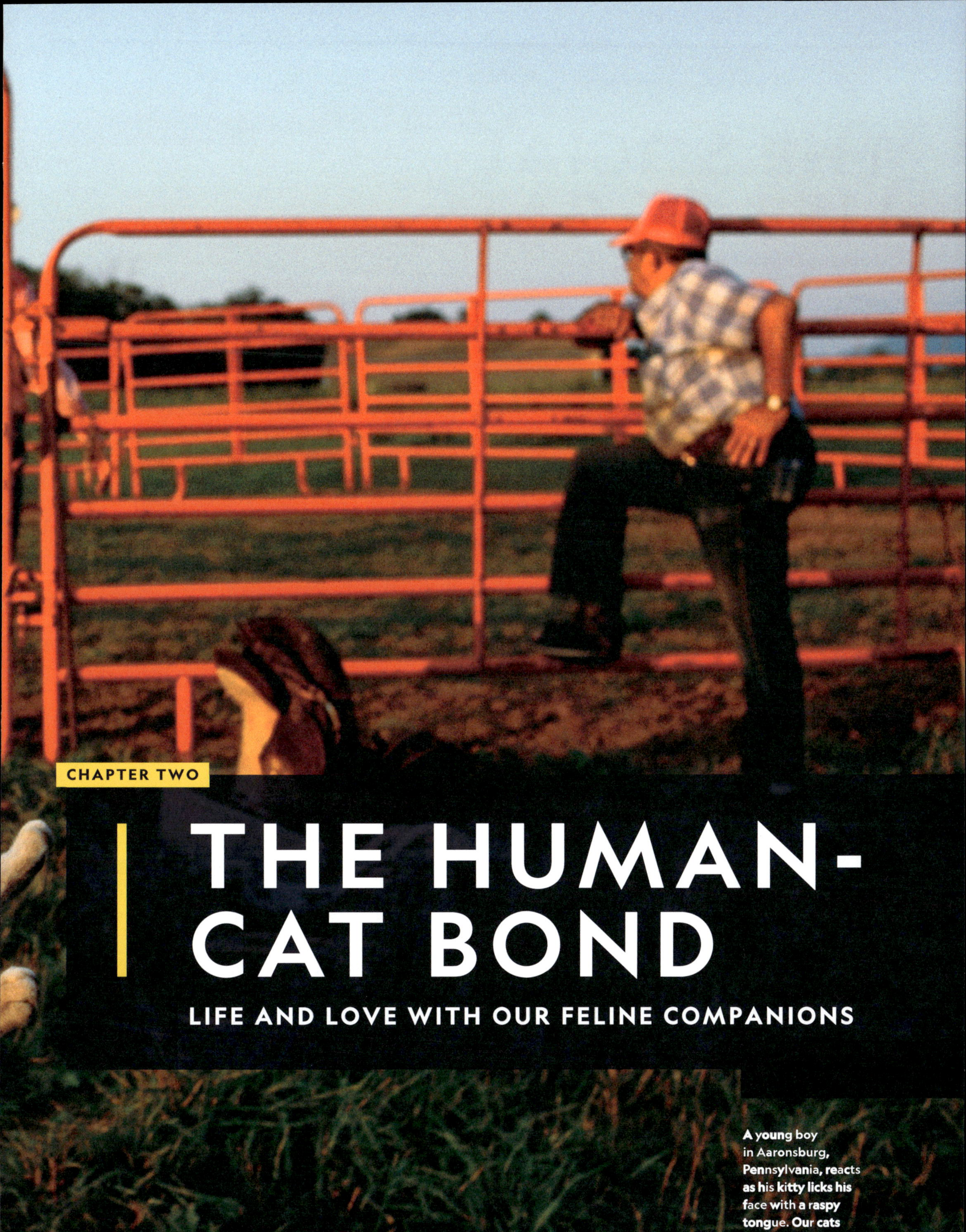

CHAPTER TWO

THE HUMAN-CAT BOND

LIFE AND LOVE WITH OUR FELINE COMPANIONS

A young boy in Aaronsburg, Pennsylvania, reacts as his kitty licks his face with a raspy tongue. Our cats

THE SOCIAL LIFE OF CATS

Group living is a relatively new phenomenon for cats. Except for lions, most species in the cat family are solitary hunters who don't live in structured social groups. Data from radio-collared wild-cats show that even their closest cousins, *Felis silvestris lybica*, the African wildcat, remain solitary.

Taking Social Cues

In their first few weeks and months of life, a kitten's social group helps teach everything they need to know about the world around them. In free-roaming cats and wildcats, kittens learn what to hunt and eat based on their mother's behavior, preferences that can last a lifetime. Veterinarians say that exposure to a wide range of food flavors and textures as young kittens may help cats develop a broader palate. Kittens also learn feline manners from the cats around them, including bite inhibition. Like human children, cats and kittens use their mouths to help explore the world around them. In this context, biting is normal. But a chomp can also

Three Catalan cat companions share a window perch and a street view in the medieval town of Pals, Spain.

INTRODUCING ANOTHER CAT INTO YOUR HOME

Thinking about adopting another cat? Talk with your vet about whether your cat would do well with another furry friend. Some cats are happiest being the only pet in a home.

When bringing a new kitty home, keep it physically separated from your original cat(s). Begin introducing the cats by allowing each to sniff a blanket or toy used by the other. Next, pet one cat and then (without washing your hands) pet the other to begin mixing their scent. After the cats adjust to this, let them see each other, while separating them by a door, stairs, or a child gate. Once the cats seem comfortable with each other, let them meet fully.

Be sure to reinforce the benefits of these visits with special treats and lots of affection. Even if your cats don't become fast friends, giving them space to call their own, including their own litter boxes and feeding dishes, stops most fights.

hurt—and cause injury. Biting another cat too firmly during play may cause the cat to stop playing. As a result, kittens learn how to soften their bite, especially with more dominant cats.

Although domestic cats may retain their ancestors' preference for solitude, kitties have also shown flexibility about their social situations. Free-living cats may be loners or live in groups called colonies. The size of a colony will vary, depending on how much food and space in a territory is available. Kittens will generally stay with the colony until they are 12 to 18 months of age, although some cats may strike out on their own earlier, and others may remain with the colony for life. More resources mean an area can support more cats.

Cat social groups are matriarchal, so colonies are generally made up of related female cats and their offspring. Kittens are often reared together, with

Cats may decide to bond with one another no matter where or how they live, or they may choose to be loners, indoors or outdoors. We may never fully understand why.

mothers taking turns nursing, hunting, and guarding the babies. Even in colonies with many members, cats continue their long history as solitary hunters. This history means that domestic cats don't have the same type of ritualized submissive signals or obvious pecking order as dogs. Instead, feline relationships are complex, being cemented with the equivalent of a secret handshake that scientists have yet to fully decode. This distinctive social structure means that a cat showing dominance in one area (such as being first to eat any scraps left by humans) may not in another (such as mating).

Any existing hierarchies frequently assert themselves at feeding time. Researchers know the most about this behavior in feral cat colonies that people feed, so it's not clear how much this occurs when humans aren't around. But individuals who regularly feed free-roaming cat colonies say it isn't unusual for larger, more dominant cats to eat first while smaller, more shy cats wait their turn for the leftovers.

Other studies of free-living cats show that aggression among toms, or intact males, is common as they will fight over access to females for breeding, to stake out their territory, and for access to food. Neutered males have smaller territories and are generally less likely to defend them. Felines' lack of a strong evolutionary history of dominance hierarchies and social interactions may mean more hissing and displays of aggression, but disputes among cats are less likely to escalate into all-out fights.

Making Feline Friends

Over the past several decades, however, researchers have begun to understand some of the basics about how domestic cats bond with each other. Cats will often form affiliative relationships with certain other cats, not unlike human friendships. These relationships are most common when cats don't need to compete for food and among closely related individuals, such as littermates or a mother and her kittens. Scientists think that this close, regular physical contact helps cats share their scent, which creates something like an olfactory friendship bracelet.

FELINE RELATIONSHIPS ARE COMPLEX, AND SCIENTISTS HAVE YET TO FULLY DECODE THEM.

The growing understanding of the importance of social bonds among cats has also revealed that a cat may be social with other felines but not with humans (such as a feral cat with limited previous contact with people), or social with humans but not with other cats (an orphaned kitten who was hand-raised by a caring human). Just as with humans, different cats will thrive on varying amounts of social interactions with their own and other species.

CAT FIGHTS AND OTHER CONFLICTS

Even in the most harmonious of homes and colonies, conflict between cats is unavoidable. A feline's long evolutionary history of solitude and territoriality makes group living challenging even under the best of circumstances. A cat's deadly weaponry (including razor-sharp teeth and paws decorated with switchblades) means that any fight between two cats could cause serious harm. Researchers think that this knowledge of mutually assured destruction helps keep some physical displays of aggression in check, because even the winner may walk away with life-threatening injuries. A fight avoided between cats, then, is a fight won.

Marking Territory

Characteristically, cats do this not through displays of dominance but by simply ignoring one another. As solitary predators with large territories, cats can detect other members of their species through scent marking without encountering them directly. These markings tell other cats that a particular spot is occupied and off-limits for hunting and mating. The core of a cat's territory is where kitty feels secure enough to sleep, eat, and be social. Beyond this hub is the area that a cat will actively defend against outsiders.

A cat, however, may roam even farther in search of food or mates. In the wild, cats may not stake out every square inch of ground into "guarded" territory, but instead may have neutral ground where they can meet and greet with no need to defend what's theirs. As long as cats have adequate food, avoiding the other cat is often the best choice. Even in multi-cat households where two felines don't get along, initial hissing and swatting are often resolved with a détente in which each side stays out of the other's way. The cats coexist by spending time in different rooms and staking out their own territory.

Drawing on the habits of their more solitary felid relatives, cats use scent to mark the boundaries of their home

Cats who are upset or frightened know how to send clear signals to all other critters: When I look and sound like this, don't mess with me.

A well-fed, well-loved cat may feel safe and secure outside knowing where home is, but most live longer, healthier lives when they don't wander.

turf. Rubbing on trees or furniture with scent glands in their paw pads, cheeks, lips, forehead, flanks, tail, and bottom, or depositing urine and feces at strategic locations are all behaviors that tell other cats, "This area is mine. Keep out." Their superb sense of smell means cats can understand these signs even from a distance.

Signs of Displeasure

Subtlety, however, doesn't always work. This is especially true during mating and when food is scarce as both hormones and hunger can make a cat bolder. Before launching into an all-out attack, a cat will display overt signs of its displeasure, ranging from hissing and growling, to attempting to make its body appear larger to deter opponents.

A CAT'S TERRITORY IS WHERE KITTY FEELS SECURE ENOUGH TO LIVE AND BE SOCIAL.

This body pose, with its characteristic arched back and fur standing on end, clearly tells passersby to stay away. Not infrequently, the other cat will retreat in a submissive posture, slinking away in a crouch with its tail held low and its ears flattened. Only after exhausting all other avenues does a cat usually follow up on physical threats.

Even in otherwise placid domestic cats, the right circumstances can trigger aggression, especially when a cat is

The fish

Cat people know that there are few clearer signals that you're loved and trusted by your kitties than when they sleep on you—sometimes, on all of you.

Though cats might sometimes startle or spook themselves, knowing their humans love them can ease many fears.

stressed or afraid. Kittens raised as singletons (without any siblings) may lack the rough-and-tumble learning from their littermates that teaches them how hard to bite without hurting their partner (known as bite inhibition) and how to read cat body language. Kittens weaned too early are also more likely to act aggressively toward other animals and with humans, drool frequently, or be quick with tooth and claw if they feel their personal space is being invaded. The stress of a vet visit can also bring out a cat's fearful body language, including hissing, scratching, biting, dilated pupils, piloerection (hair standing on end), or arched back.

For some cats, even the nearby sight or smell of another feline is enough to trigger stress-based behaviors that sometimes lead to aggression. This animosity can be shown by an increase in territory guarding at doors and windows, or through less desirable behaviors, such as urinating and defecating outside the litter box. Whether a cat is too intimidated to visit its standard latrine, attempting to show the other cat that its home territory is already occupied, or something else entirely, changes in litter box habits are the number one cause of vet visits and for animal surrenders to shelters. A veterinary workup can rule out medical causes for these changes, such as a urinary tract infection or blockage. In the meantime, humans can help their cats de-stress.

Humans don't always fit nicely into this feline world. If cats don't experience human contact (such as petting, feeding, cuddling, and being picked

up) as young kittens, many will spend the rest of their lives avoiding us. Biologists have found that this isn't rooted in a dislike for humans but rather in a fear response that remains stamped into a cat's brain, even after lots of positive human interactions.

Humans are predator-size, and it's not hard to see how a cat might think that our bumbling efforts at showing affection can mimic what might happen shortly before kitty becomes dinner. These feral instincts are easiest to overcome with extensive socialization before a kitten is 12 weeks old. But with hard work and patience, even some lifelong feral cats can adapt successfully and happily to life with humans.

BEST FURRY FRIENDS FOREVER

Despite their reputations as loners, cats are capable of forming close relationships with members of their own species, with people, and even with dogs. Cats might not be as dependent on social interactions as dogs are, but that doesn't mean they are always happy to live solo, either.

Domestic cats first began socializing regularly during the rodent boom that accompanied the development of agriculture. For the first time, cats didn't have to compete with one another for food and shelter. As a result, cat social relationships are built on a foundation of adequate resources. Facilitating cat social relationships, such as having several cats in a home, means ensuring the cats have ample food, litter boxes, places to sleep, and space for "me" time as well. Veterinarians recommend having a litter box for each cat in the house, plus one extra (if your home has three cats, then you want four litter boxes). It may take some trial and error to figure out your cat's preferences for types of litter and boxes, but this will also prevent elimination problems in your home.

ONE OF THE CLEAREST SIGNS OF A CLOSE CAT RELATIONSHIP IS PHYSICAL CONTACT.

Love and Affection

Cats that have lived together since birth generally get along well with one another, but just as with human siblings, friction can develop as cats get older. Two adult cats can also meet and become closely bonded. One of the clearest signs of this close relationship is physical contact. Cats in a bonded pair will often sleep curled up next to one another, and rub against each other in greeting. Paired cats also groom each other, especially around the face and neck. This offers the double advantage of cleaning hard-to-lick parts of the body while also sharing scent from glands in that area. Their mutual grooming means that closely bonded cats will have a "we" instead of just a "me" scent.

This affection and what psychologists call cathecting, or bonding, runs deeper than even the biggest ailurophile is often aware. Scientists and cat lovers have documented many cases in which multiple female cats in a feral cat colony who have kittens at the same time will often share nursing, hunting, and babysitting duties. These queens subsequently

Sometimes that paw touch might be loving, sometimes it might be for marking or grooming, and sometimes it could just be a playful bop on the head.

raise their kittens as one large group. Mama cats will even adopt and suckle orphaned kittens, readily adding the extra mouth to feed to their own broods. Cat rescues may take advantage of this propensity to nurture other kittens so that orphaned or abandoned kittens can get the social and biological benefits of a feline mother.

Studies have shown it's not because the mother cat doesn't know how many kittens she has or that a newcomer isn't biologically "hers," but that her maternal instincts overpower any hesitation she has. As long as a rescue organization is feeding mama ample calories to produce milk, and both kittens and nursing mom are checked for parasites and signs of contagious disease, the benefits of surrogate nursing can be tremendous.

All in the Family

Growing awareness of the importance of friendship in cat well-being means that more and more rescue organizations are asking people to adopt kittens in pairs or have another young cat at home to offer suitable companionship. Experts say that adopting two (or three!) kittens together does more than just increase the cuteness factor in the household. It can mean not only fewer behavioral problems, better grooming, increased mental stimulation, less pickiness around food, and more exercise for the cats but also the warm feeling that we humans get from saving two lives instead of just one.

If these signs of cat bonding seem familiar to those of us lucky enough to share our lives with a cat, that's because they are. Cats show affection to their humans in almost the same way as they do their feline friends, which supports the idea that cats see us as just another cat. Everything from purring, twining themselves around our legs, giving us baths with their sandpaper tongues, kneading their paws (what the cat-savvy call "making biscuits"), and even rolling over on their backs to reveal their soft and vulnerable bellies are all behaviors that cats use to show affection for each other. Even their "gifts" of decapitated birds and eviscerated rodents are probably an effort to share food with a companion who isn't a good hunter. Cats are obviously smart enough to know that humans are not as small, furry, and agile as a bona fide feline, but they generally treat us like

Kittens in a bonded pair (left) are each other's young companions. They also help each other learn how to be adult cats who live happily with humans and other animals.

Growing up with a cat helps kids learn about trust, empathy, and responsibility.

we're one of the gang. Humans can show their affection by not only giving pets treats but also blinking their eyes slowly, which likely translates to "I love you" in cat.

When cats do form friendships with humans, the affection they feel can be deep and lasting. Although our feline companions may not show love by wagging tails or greeting us at the door, scientists have found that cats see their humans as a source of love, comfort, and security. Tests in controlled settings show that two-thirds of cats will turn to their human companions to evaluate whether the situation is safe and to get a boost of affection and confidence before venturing off to explore.

HOW TO SPEAK CAT

Just because cats can't talk doesn't mean they can't communicate effectively, both with humans and with their own species. Yowls and hisses aside, cats rely primarily on body language and scent to say everything from "This turf is mine" to "Go away" to "Let's play."

Talking Scents

A cat's exceptional sense of smell means that our furry friends live in a rich scent-scape we humans and our so-so sniffers can't fully imagine. A variety of scent glands are distributed across a cat's body, from their chin and cheeks to their paw pads, flanks, anus, and tail. When they rub those glands on an object, animal, or person, they transfer that scent. Only other cats can detect these special pheromones. The precise mixture of chemicals deposited by each cat carries a wealth of information about their age, sex, mating status, and general health.

Urine marking and spraying, especially for indoor cats, are more overt and potentially problematic and also most common in unneutered males. A cat sprays by backing up to a vertical surface and depositing a small amount of urine. More rarely, a cat will spray on a horizontal surface. Just like scent glands, this sprayed urine contains pheromones that tell of a cat's availability for mating and that it occupies a specific piece of turf.

Because of the close links between scent and territory, cats will often rub or spray most frequently around entries and exits (like doors and windows), and around the edge of their home turf. Cats will also mark new items in their territory, such as furniture, freshly planted trees, and even human guests and visitors.

Because cats use scent to signal "ownership" of a person or place, a cat

Scents help cats and kittens navigate the world. It is crucial for how they communicate with one another.

TALK TO THE TAIL

Scent, however effective, can't communicate everything. For conveying much of what else they have to say, cats use body language. One of the most expressive parts of a cat's body is her tail. Learning how to read a cat's body language can help us be better companions to our favorite felines.

An upright tail, often with the tip curved to one side, means cats are feeling relaxed and friendly. Similar to dogs, scared cats will hold their tails down close to the ground or between their legs, slinking away in retreat with their legs crouched and abdomen close to the ground, staying close to protective cover for safety. A fearful kitty will make her fur stand on end (known as piloerection) and arch her back to look as big as possible. A slowly swishing tail can mean the cat is feeling curious or playful, but a lashing or whipping tail means the cat is feeling fearful or aggressive.

who is stressed, anxious, or threatened may increase this marking behavior. The "self" smell may also explain why some cats in shelters will try to hide by sleeping in their own (sometimes used) litter box. In such a strange and overwhelming environment, the litter box is one place these cats know is definitively theirs. If you find your cat spraying indoors, veterinarians recommend ruling out potential medical problems and then developing a plan to help reduce or eliminate this behavior. Neutering is often the first-line treatment for this problem. But for fixed cats, behaviorists recommend thoroughly cleaning the sprayed area with a pet-friendly enzymatic cleaner to remove all trace of the scent. Lingering odors may encourage the cat to continue spraying.

The next step is to identify what is triggering the spraying behavior. If the source of stress can't be pinpointed or altered (for example, with a new baby in the house), a diffuser with calming pheromones for cats may help them relax. Blocking a cat's access to spots where they like to spray, such as by closing doors or placing large objects in the way, may also help prevent spraying and urinating outside of the litter box. If the problem persists, antianxiety and antidepressant medications like clomipramine and fluoxetine have been shown to help.

What's in a Meow?

Cats vocalize to humans in much the same way they do with other cats—with one main exception. The plaintive meow that grabs our attention is rarely used to speak with other cats. Although cats may growl, howl, yowl, and hiss to communicate with other members of their own species, meowing is almost nonexistent after kittenhood. That particular vocalization is unique to cat-human communication. House cats, scientists have found, meow far more frequently than feral cats.

So where did the cat's meow come from?

BUTT WHY?

Anyone who has shared their home with a cat has greeted their friendly feline only to have kitty do an abrupt 180-degree turn and shove their butt in their human's face. Cat behavior experts say that this really isn't a sign of feline derision or of showing off those hours of careful grooming of their nether regions.

Instead, your cat is just saying hello. Presenting their hindquarters makes cats vulnerable, but it also allows them to share their scent more easily with other creatures. Each cat's butt has a signature scent, which is distinctive and unique to them. When they show us their butts and present themselves to us in this open way, cats are greeting us, their humans, as a sign of friendship. They're also showing us that they aren't a threat to us. The butt-first greeting, then, can be a sign of their affection and trust.

If you were this tiny and helpless, you might be crying out for attention, too.

One theory suggests that a meow shares many of the same acoustic qualities as a baby's cry. With humans hardwired to respond to a baby, it wasn't much of a stretch to respond to a cat. Cats, for their part, may have simply learned that certain behaviors—meowing included—get attention from humans. Those cats that meowed more often as adults may have gotten more human handouts, making them more likely to survive and raise more chatty kittens just like themselves.

All meows, however, are not created equal. Even novice cat lovers can tell the difference between an angry howl telling others to stay away and a sweet mew asking for a scratch behind the ears. In 1895, when cats were not as commonly household pets, a British scientist wrote an essay identifying 17 different meows and what they meant. Now, however, scientists know that a cat doesn't have a specific meow that means "fur" or "cooked meat." In fact, a cat's vocalizations don't qualify as true language, because cats don't know the meaning of their own meows. What they do know is that a meow will cause humans to pay attention to their demands and, more often than not, get what kitty is looking for.

More rigorous studies have found that humans can learn the subtle differences between a kitty meowing because it's hungry, scared, playful, or just bored. And humans do best when interpreting the vocalizations of their own cats compared with those of a cat they've just met. Research suggests that cats learn to make these different

SCIENTISTS ARE NOW RESEARCHING CAT MEOWS TO SEE IF THEY HAVE DIALECTS SIMILAR TO THOSE OF HUMANS.

meows at different times based on how humans respond to their cries. Swedish scientists are currently investigating whether cat meows have dialects similar to those of humans. Ten thousand years of domestication have given cats plenty of chances to figure out exactly how to push our buttons and to convince us to do what they want.

Whatever your cats are trying to tell you, talking back to your four-legged companions may make them even chattier. Experts recommend speaking to your kitty in a soft and calm voice, as cats are sensitive to vocal tone, even if they don't know exactly what you're saying. One thing cats do likely recognize would of course be their own names. If they don't come when called (and most don't!), it's because they are deliberately ignoring you, not because they don't know you're calling. Some cat lovers say that the shake of a treat bag or a crinkly toy is a more reliable way to get a cat's attention.

A regular at Bistrot Le Louis IX in Paris's Île de la Cité, Caramel the cat keeps another customer company.

NAVIGATING LIFE WITH HUMANS

Dogs were one of the first animals domesticated by humans. Thousands of years of this coexistence have helped facilitate human-dog communication. Our reliance on dogs for tasks like hunting and herding has meant that we have already smoothed over many of the communication snafus that can pop up between people and four-legged friends.

Living Together

Cats, however, have a much shorter domestication history—and an incomplete one at that. This, combined with a natural independent streak, means that cats aren't born knowing how to live with humans. Instead, each kitten needs to learn how to share its world with the affectionate but clumsy humans who love it. The good news is that cats of any age can learn to be more trusting of humans, although this task is quickest and easiest with young kittens.

Being born with several littermates and a mother means that kittens are

CATS OF ANY AGE CAN LEARN TO BE MORE TRUSTING OF HUMANS.

Sometimes our cats remind us that they want to play and that they want us to play with *them*. Preferably now.

EXPANDING YOUR CAT'S WORLD

Pushing a cat's boundaries in a loving way is hard work, but the key is to make these new experiences pleasant and fun from the very beginning.

To help your cats adapt more easily to new circumstances, including those that might be difficult or seem strange for them at first, try giving them their favorite small snacks with situations that may make them stressed or afraid. These might include going in the cat carrier, riding in the car, or being examined by a vet.

Create a positive association for your cats between these new experiences and old favorite treats, plus petting and praise. You can also try offering toys to foster and build pleasant emotions and associations between what's familiar and what's new.

Keep "training" sessions short to avoid overwhelming your cat. Remember that this training is only valuable to you all if it's also fun.

naturally socialized to be with other cats. During the socialization period of development, a kitten learns how to hunt and how to get along with other cats (whether or not they choose to get along is up to them). This happens largely through physical play that begins between two and seven weeks of age, once kittens are a bit more mobile. By four weeks, kittens begin playing with their siblings. Between four and five weeks, kittens begin batting at toys and stalking and pouncing on objects. Play peaks at around 16 weeks, though adequate playtime remains important for cats of any age.

Although life in a feral cat colony can be tough, group living offers kittens ample playmates and other watchful adults who can teach even the most rambunctious kitten how to be a cat. So-called "kitten kindergarten" classes (in which a handful of humans bring their cats together for regular play-dates) can help rescue cats gain the benefits of being in large social groups without risking their lives outdoors.

Kittens take many of their cues from their mother. The situations where she is most comfortable rub off on her young. Kittens born to a feral mom learn from the beginning that humans are dangerous and best avoided. Even youngsters born to the friendliest mama cat will grow up afraid of humans if they don't experience regular human contact by

Part of becoming a sociable cat is just learning how to let other kitties get close, and to feel safe and secure while they do.

CATS FEEL MOST AT HOME IN PLACES THAT REMIND THEM OF WHEN THEY WERE YOUNG.

the time they're seven weeks old. As little as five minutes of human interaction a day before seven weeks of age—and ideally starting at two weeks—can make a lifetime of difference.

Experts in feral cat behavior say that the best way to help feral kittens adapt to live indoors with humans is to pair human presence (something scary) with special food that gets a positive response. Over time, the kittens will associate humans with tasty snacks and realize humans aren't that dangerous. These early exposures also help create a more secure bond between cats and humans. Researchers found that two-thirds of both adult cats and kittens look to their humans as a source of security and comfort. A third remain more aloof—a proportion that mirrors attachment styles in humans.

Feeling at Home

Socialization isn't just about getting along with other cats and with humans. It's also about localization, or being comfortable in different environments. Starting life beneath a shed creates a very different set of experiences than being born inside a comfortable, bustling home with a human family. Cats generally feel most at home in places that remind them of when they were young. Early localization helps expand their

"LOST" AND FOUND: CATS, TERRITORY, AND NAVIGATION

Around 15 percent of all pet cats sometimes escape to the outdoors, with three-quarters eventually finding their way home. But scientists have wondered how cats who have never ventured outside still have a near-miraculous ability to navigate—sometimes across hundreds of miles.

While salmon use Earth's magnetic field, dung beetles use the Milky Way's path, homing pigeons use low-frequency sound waves, and migratory birds in flight use visual clues from far below, cats use scent to mark home territory, but it's unclear what guides a cat over longer routes.

Whatever drives it, a cat's homing instinct is strong. In one study, a mother cat found her kittens after being separated from them by up to four miles (6 km). But rather than relying on kitty's built-in GPS, veterinarians recommend that all cats be microchipped and wear a breakaway collar with an ID tag, to make it more likely you'll be reunited.

Weaning a kitten from bottle-feeding (opposite) usually takes four to six weeks, or about as long as if he were being fed by his mom. Most kittens are completely weaned by 10 weeks.

Even an occasional friendly visit with humans can make a big difference in how a cat begins to see us.

comfort zone and makes it more likely they will live well with humans in the great indoors. And though jokes about cats hating trips to the vet and having their nails trimmed aren't about to stop any time soon, getting cats familiar with these activities early in life can make veterinary care less traumatic later on.

But just because a cat's early years weren't ideal doesn't mean kitty won't be a good pet. Cat rescue groups have been doing pioneering work with free-roaming and feral cats to socialize them to humans. Although not all of these cats indicate they would be happy living with us, a surprising number seem willing to change their minds about people and adapt well. However,

it takes a lot more time and patience to help these kitties adjust to living with creatures they once feared.

The reason, according to animal behaviorists, is known as social referencing. A trusting cat used to being around humans requires only a few positive interactions with a stranger before beginning to show friendliness. It also takes major negative experiences to override that cat's early life learning that humans can be trusted. A less-socialized or more feral cat has a negative social reaction for humans. These cats need far more positive interactions with strangers to establish trust and overcome their belief that humans are threatening and dangerous. Any negative experiences quickly overcome the new, positive interactions. In other words, kitten socialization to humans offers a positive reference base a cat can build on throughout its life. Abuse or a lack of positive social experiences with humans creates a negative reference history that will color a cat's later interactions.

THIS IS YOUR CAT ON *TOXO*

Toxoplasma gondii may be just a single cell, but the parasite's impact on behavior can be profound. This distant cousin of malaria can survive in almost any mammal, but it can only sexually reproduce in cat intestines. To ensure this happens, *Toxo* attracts rodents to the smell of cat urine—something they normally avoid. This attraction makes it more likely the rodents will eat the *Toxo* oocysts shed by the cat, which completes the parasite's life cycle.

Healthy animals with *Toxo* develop only mild, flu-like symptoms, though veterinarians report eye issues in *Toxo*-infected cats. And pregnant women are advised to avoid cleaning litter boxes, because infection can cause severe neurological issues in developing fetuses.

Research notes increased risk of schizophrenia and other psychiatric conditions, including heightened risk-taking and potentially self-destructive behaviors, in people infected with *Toxo*. However, these findings remain controversial, and scientists still aren't sure of *Toxo*'s full impact on humans.

HOW CATS LEARN

Once they've gotten their cat to use the litter box consistently, many cat people consider their pet training complete. But a growing body of research says that cats are more trainable than people first thought and this type of rich human interaction promises to make their lives (and yours!) more fulfilling.

Finding What Motivates

The most important aspects of training your cat, behaviorists say, are finding out what motivates your cat, and using that to make the training fun. Whereas most dogs are willing to perform any number of tricks to please their people (although a few treats never hurt), human affection isn't as innately rewarding for all cats as it is for dogs. A cat's independence can also make your cat less inclined to follow commands. Instead, helping cats learn basic skills such as coming when called or holding still for grooming, as well as simple tricks, such as shaking paws and jumping through hoops, usually relies more on special treats that reward them for doing what you want them to.

Many pet trainers rely on clicker training, which uses a small handheld device to acknowledge a desired behavior with an audible "click." Pairing the click with a treat creates an equation that tells kitty BEHAVIOR = CLICK = YUM. Combine the clicks and treats with a calm, upbeat voice and lots of praise when they do as prompted.

To teach a complex series of behaviors, such as jumping through a hoop, start simple with easily learned parts of the behavior. For instance, use a treat to lure the cat to approach a hoop that's touching the ground, and reward her for that. Next, lure her into stepping through the hoop. Then move on to raising the hoop inch by inch so the cat must jump to go through it.

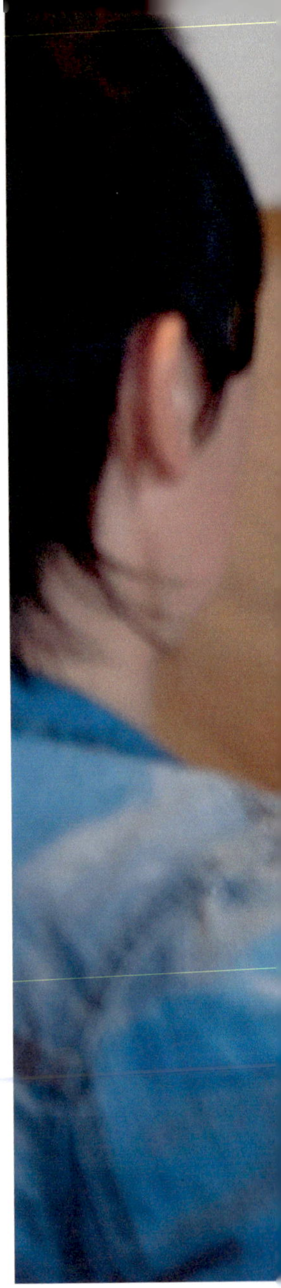

Training a kitten to use a scratching post (left) is easier with catnip or a pheromone product, reinforced with yummy treats for learning good nail grooming behavior.

These Norwegian Forest cats wait for more of the tasty hair ball remedy their human gives them as a reward after their comb-outs.

Avoiding What Doesn't Work

Punishing cats for bad behavior, such as urinating and defecating outside the litter box or scratching the furniture, doesn't work. Instead, veterinarians recommend trying to figure out what's driving the undesired behavior. Inappropriate litter box usage could be caused by anything from stress over a new baby to pain from a urinary tract infection. Rather than raising your voice if your cat scratches the couch, encourage alternate behaviors, such as the use of a scratching post. Your cat won't understand why you are angry, and, over time, will begin to respond to you with fear, rather than joy.

USE WHAT MOTIVATES YOUR CAT TO MAKE TRAINING FUN.

CRACKING THE CODE OF CAT EMOTIONS

A cat may never give us puppy dog eyes or look guilty after eating a forbidden snack. Because scientists were looking at cat communication through canine lenses, their lack of doggy-style expression led some to believe that cats didn't connect much to humans. Newer studies, however, show that although a cat's communication style is different and more subtle than a dog's, it's no less rich and nuanced. With a bit of effort, every ailurophile can learn what their cat is saying, even if kitty never utters a word.

Feline Feelings

Behavioral biologists from the University of Guelph in Canada have found that a small subset of humans are unusually adept at reading feline emotion. The scientists asked internet users (aka the cat's biggest fan club) to watch 20 short videos of feline expressions and rate whether the cats were distressed or happy. The average user got 11.85 out of 20 correct—not much better than chance. But a small subset of humans, around 13 percent of test takers, got at least 15 out of 20 correct. These individuals were disproportionately

When you try to read your cat's behavior or body language, what she's feeling may seem complex, even mysterious. Or she may just want a nap.

DECODING YOUR CAT'S MOOD

ALTHOUGH A CAT'S face may never be as expressive as a dog's, humans can learn to read a cat's body language to understand their mood.

Relaxed
Purring; ears up and forward; eyes partly closed; very happy kitties may knead their paws as well.

Angry or aggressive
Growling or yowling; ears back; pupils constricted; fur standing on end; swishing tail (the faster the twitch, the angrier the cat).

Playful
Alert and attentive; whiskers forward; ears swiveling; pupils partly dilated; tail up. Essentially hunting behavior: to cats, playing and hunting are the same.

Frightened
Hissing or growling; whiskers and ears back; pupils dilated; tail between the legs; back arched; fur standing on end (think Halloween cat).

Nervous
Pupils dilated; ears sideways and/or back; tail held low; walking close to the floor.

Loving and trusting
Slowly blinking eyes; gentle head butts and cheek rubbing; rolled over with belly exposed (not necessarily an invitation to pet that tummy, but a sign of trust).

young and female and had professional veterinary experience. Researchers suspect that this expertise helped these cat-emotion detection superstars shine. By spending lots of time with cats, some people have developed a keen eye for understanding their behavior.

Experts also caution against using one small snippet of information to identify a cat's emotional state. Many cat lovers know relatively little about cat body language, and a cat's overall demeanor and well-being are far more complex than a feeling at a given moment. It's why cat behavior experts recommend starting with the big picture.

IS YOUR CAT IN PAIN? HERE'S HOW TO TELL

Despite their reputation as fearsome predators, many cat species are also small enough to be prey. This gives them a strong incentive (and a hardwired instinct) to avoid showing they are in pain. As a result, it can be hard for even vigilant cat lovers to tell when kitty is hurting. Here are some behavioral changes that may mean a trip to the vet is in order:

- **Hiding; withdrawn, quiet, or grumpy mood**
- **Loss of interest in people and activities; less affectionate; avoids petting, being handled, or playing**
- **Restlessness, trembling, agitation, crying, yowling**
- **Hissing, growling, and other aggressive behavior**
- **Closed eyes; excessive blinking**
- **Tail flicking**
- **Changes in litter box habits**
- **Oversleeping**
- **Lack of grooming or excessive licking of body parts**
- **Reluctance to move or walk**
- **Avoidance of bright light**
- **Change in gait, posture, or movement; limping, difficulty jumping**
- **Decrease in appetite**
- **Rapid or shallow breathing**
- **Flattened ears, narrowed eyes, and upturned whiskers**

WITH EFFORT, EVERY CAT LOVER CAN LEARN WHAT THEIR CAT IS SAYING.

The key is putting yourself in your cat's paws and imagining how the world might appear to them. Start with their environment: Is there something (or someone) making your cat feel threatened? If so, the upright tail might be your their attempt to look larger, rather than a sign of content-ment. As animals that have evolved as both predator and prey, cats have the ability to look smaller when threatened or to appear bigger and more threaten-ing to frighten away potential rivals and predators.

Also note how your cat orients his body. Like humans, cats typically look in the direction they want to go. If your cat is facing sideways to you and eyeing the door, he could be feeling shy and wanting to make a quick getaway. It also gives him a lead in case he gets chased. Ditto for hunkering down in a crouch. But body position isn't always hard to interpret. A cat rubbing his head against your ankles or body likely wants to be approached and wouldn't mind a scratch or two behind the ears, thank you very much.

Cats may hide when they feel sick or threatened. They instinctively know that, in the wild, a vulnerable animal is a target for predators.

CHAPTER THREE

CATS IN OUR HISTORY

A HUNDRED CENTURIES OF LIVING WITH HUMANS

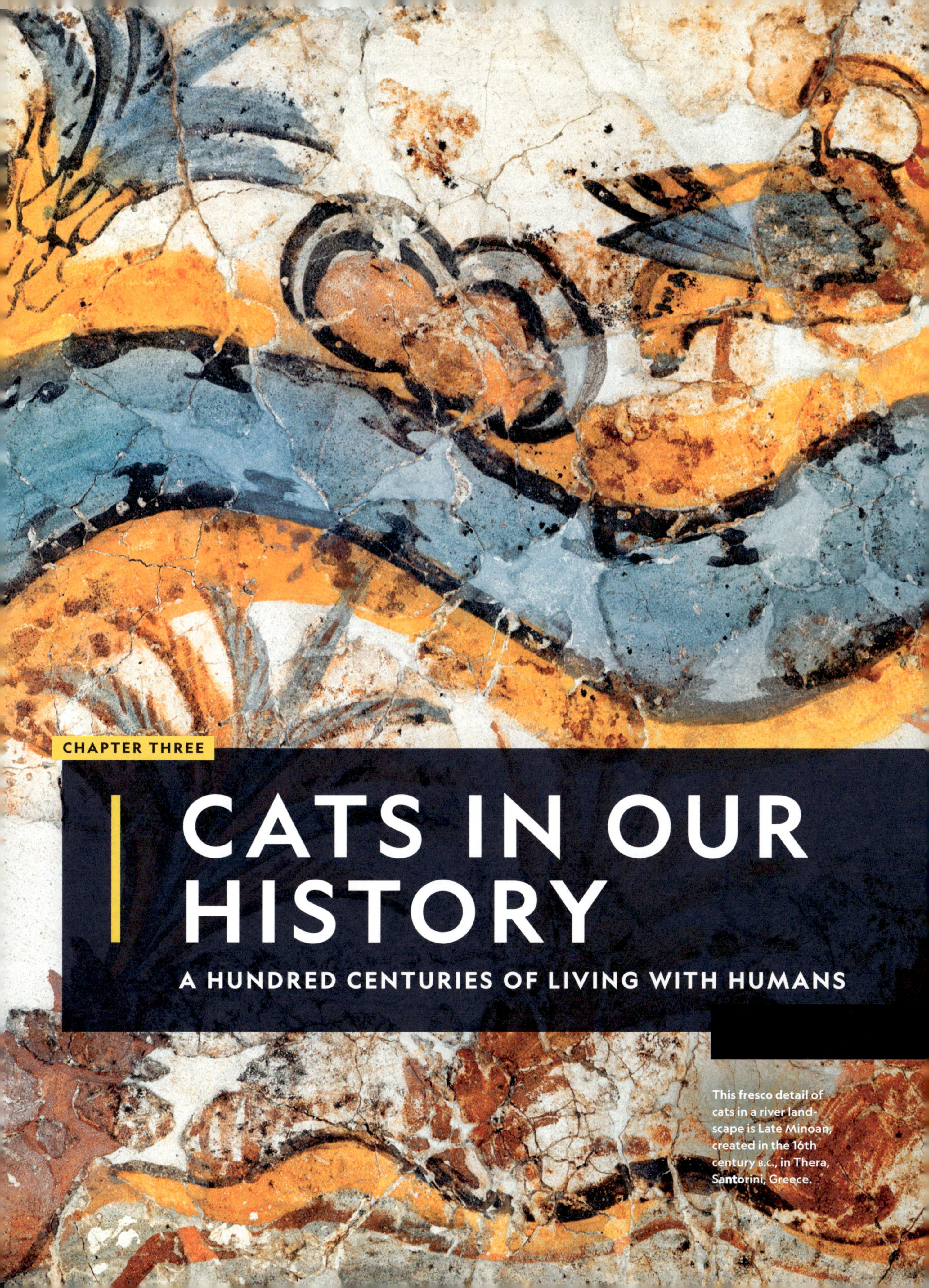

This fresco detail of cats in a river landscape is Late Minoan, created in the 16th century B.C., in Thera, Santorini, Greece.

A TIMELINE OF CATS

Cats were sacred in ancient Egypt. Mummified cats were dedicated to the cat goddess Bastet or entombed with their humans so they could be together in the afterlife.

For many years, researchers believed that the ancient Egyptians were the first to begin keeping cats as pets around 4,000 years ago. But more recent archaeological and genetic discoveries show that cat domestication began around 10,000 years ago.

A grave in the Neolithic village of Shillourokambos, in what is now Cyprus, contains the first evidence that humans saw cats as something more than mobile rodent hunters. A team of French archaeologists found a human and an eight-month-old cat buried just 16 inches (40 cm) apart, in a grave dating back 9,500 years that also contained offerings of seashells, stone tools, and ocher. Because cats aren't native to the island, researchers believe they were brought over by boat, either deliberately or as stow-aways, perhaps with settlers from the Middle East. This indicates that humans and cats may have been living together for at least one hundred centuries—more than twice as long as previously thought.

CATS HAVE BEEN WITH US MUCH LONGER THAN WAS ONCE BELIEVED.

UNSINKABLE SAM

Cats proved as useful on naval vessels as they did on merchant ships. During World War II, British sailors supposedly rescued Oscar, a black-and-white tuxedo cat, who was found floating among the wreckage of the German warship *Bismarck* in May 1941. He was named Oscar (sometimes spelled Oskar for his German roots) from the International Code of Signals for the letter O, or "man overboard." The crew of the HMS *Cossack* adopted the waterlogged moggie as the ship's mascot. But the *Cossack* was torpedoed in October 1941, and the kitty was brought on board HMS *Ark Royal*, an aircraft carrier. This, too, was torpedoed, and Oscar became known as Unsinkable Sam. He is the only known mariner to have served with both the Royal Navy and the Kriegsmarine. The kitty eventually retired from ship life and spent the rest of his days as a landlubber in Belfast, Northern Ireland.

An archaeologist works inside the tomb of Khufu-Imhat, one of the seven tombs belonging to the New and Old Kingdoms of Egypt that were discovered at Saqqara necropolis in 2018. Dozens of cat mummies, hundreds of wooden statues of cats, and a bronze figure dedicated to Bastet were among the finds.

The First Domesticated Cats?

A gap remains in our understanding about what happened in the cat world during the 5,500 years between the Cypriot find and the earliest art depicting domestic cats in Egypt. Archaeologists found a cat tooth dating to around 9,000 years ago in Israel, and another dating to about 4,000 years ago in Pakistan.

Scientists also found evidence of cats from 5,300 years ago in the ancient agricultural village of Quanhucun in Shaanxi Province, China, although it isn't clear whether these were actual domestic cats or Asian wildcats that happened to be living nearby. One clue to domestication from the Quanhucun site is the remains of an elderly cat,

were actually leopard cats *(Prionailurus bengalensis),* not true house cats. If leopard cats were ever domesticated, it didn't last; all modern domestic cats in China are descended from Middle Eastern cats.

From Hunters for Hire to Lap Cats

Even if they didn't start the trend, the ancient Egyptians helped to transform cats from mobile grain protectors to beloved pets. Feline skill in killing rodents and snakes endeared our furry friends to the Egyptians. They had several cat deities, including the goddess Bastet, daughter of Ra, the sun god. The ancient Egyptians mummified some of their cats (though detailed x-ray analysis of some mummified felines shows that local wildcats were also included). Archaeologists have even found vast necropolises with a variety of animals. Egyptians were reported to shave their eyebrows in mourning after one of their cats died.

Although cats may have reached their ancient high-water mark of status in Egypt, other cultures also found that felines were useful to have around. Wherever they were, a cat's rodent-hunting abilities proved as

which showed that it had successfully aged there. Other remains found at the site are of a cat who had eaten millet, which suggests that the cat scavenged food from humans or had been fed by them. Although some scientists claimed that the Quanhucun site was evidence of early cat domestication in China, more recent analysis shows that these

This sculpture of a reclining tiger made during China's Song dynasty (right) dates from the 11th century.

popular in China as it did in Egypt. The Chinese had their own cat goddess, Li Shou, who they believed promoted pest control and protected fertility. An ancient Chinese myth held that the gods originally nominated cats to run the world, but when the cats were found asleep on the job, the gods turned the task over to humans. During the Song dynasty from A.D. 960 to 1279, the Chinese valued cats not only for their rat-catching abilities but also for their appearance, and they were popular pets.

From Hunting Mice to Hunting Souls

Silkworm-eating mice in Japan led to a close relationship with felines there, too. Making the journey across the East China Sea and the Sea of Japan in Chinese ships, cats arrived around A.D. 500 in Japan, where they soon found a special place in society. The Japanese were beset by mice that ate their valuable silkworms, so the newly imported cats offered valuable rodent control. Several ancient shrines were built to honor cats, including one on the small island of Tashirojima, off Japan's east coast. Hundreds of stray cats now call "Cat Island" home.

In medieval Europe, however, the cat's aloof nature stood in direct contrast to its natural usefulness as a mobile mousetrap. Despite the cat's seeming domestic status, humans couldn't control cats in the same way as dogs and livestock. The Catholic Church began to draw parallels between cats hunting mice and the devil hunting souls. In 1483, Pope Innocent VIII declared that the cat was the favorite animal of the devil—and witches. Just as witches and heretics were burned at the stake, so, too, did countless cats meet a fiery end. Black cats were targeted above and beyond their multicolored counterparts. Despite centuries of atrocities, however, felines survived in Europe and elsewhere.

THE FIRST CAT PHOTOS TO GO VIRAL

Long before cats took over the internet—and long before there was an internet to conquer—renowned photographer and color photography pioneer Willard Culver spent months taking elaborate staged photos of some highly pampered pussycats.

For Culver, the series was as much about groundbreaking photography as it was about capturing timeless images of beloved cats. He took his color prints with a synchronized photoflash. The feline photo series was the first time such photographic images were ever published. Culver's secret to the perfect pet portrait?

"With a bit of salmon or catnip, sometimes liver, approximate friendship could be developed, though many times short-lived. A false move on my part would change the most polite and complacent pussycat instantly into a defiant and combative creature that revealed the savagery of the jungle."

Culver's series of photos graced an article written by Frederick Eddy in the November 1938 issue of *National Geographic* magazine.

NEWLY IMPORTED CATS OFFERED VALUABLE RODENT CONTROL.

Have Cats, Will Travel

Global DNA studies of cats show that our four-legged friends sailed with humans as they moved around the world. While on board, cats helped protect the limited food supply from mice and rats. They also gave sailors companionship. In time, it became bad luck to sail without a cat on board. When Europeans sailed to the Americas, they brought cats with them. Whether left deliberately or accidentally, domestic cats soon populated every port city. Between maritime movements and the spread of agriculture, it wasn't long before domestic cats had made the world their own.

By the 1800s, European attitudes toward cats had shifted again. This time, instead of viewing them as furry rodent control or agents of Satan, humans began to exploit a cat's ability to become a pampered pussycat. Some enterprising humans began breeding cats much like humans had long bred dogs. But instead of breeding a cat to accomplish a specific task, such as herding or guarding, cats were bred largely for their appearance. Persians, noted for their luxurious fur and affectionate nature, became popular in upper-class circles. Most cats, however, remained the equivalent of beloved mutts, roaming outdoors and reproducing when, where, and with whom they preferred. This independent lifestyle persisted up to the 1950s, when the invention of a key product began to change everything.

A detail from this 15th-century book of hours shows a cat catching a mouse, helping stop the spread of plague. The Black Death killed 200 million people in the 1300s.

A HALF-CENTURY OF CHANGE

Up until the mid-20th century, going outside was a necessity for cats. In the days before commercially available cat litter, outdoors was the only place cats could do their business. Although some cat people began creating their own indoor cat toilets using a box filled with any combination of sand, sawdust, paper, and fireplace ashes, these ersatz litter boxes did little to eliminate the pungent odor of cat urine.

All that changed in January 1947. Michigander Edward Lowe worked selling clay, sawdust, and sand to soak up industrial spills. A neighbor, tired of her cat tracking ashes around the house, asked Lowe for something different. He gave her a bag of clay granules, and the rest, as they say, was history. Lowe's invention, which he called Kitty Litter, spawned a $10 billion global industry that projects 4 percent annual growth between 2022 and 2028.

Lowe's genius wasn't so much the invention of cat litter (he simply repurposed an existing compound) but in marketing it. He highlighted clay's ability to reduce odor, along with a litter box as a convenient place for urine and feces. Compared with ashes, clay was less likely to be spread around the house by little

THE TRUTH ABOUT CAT PEOPLE AND DOG PEOPLE

According to the internet, the world is filled with either cat people or dog people. And though many people may have a preference for the type of furry companion they live with, some psychologists believe the divide goes far deeper than that.

A 2010 study of self-identified cat and dog people found that dog people were more outgoing and had fewer negative emotions. An independent follow-up study found similar results. Whereas dog people seemed to be more outgoing, those who preferred cats were more sensitive, open-minded, and nonconformist, which reflects many of the same stereotypes about the animals themselves.

But not everyone is convinced there really is a difference between these groups of animal lovers. Some sociologists believe that any issue about cat people versus dog people is just a proxy for feminine versus masculine. What's more, plenty of people like both cats and dogs, and an even smaller group aren't really into pets at all.

Edward Lowe expanded his Kitty Litter empire into a full range of cat care products, such as those in this 1955 display case.

feet. What's more, clay clumped, making it easy to remove used litter, cutting odors more and creating a tidy spot for kitty to do its business. By 1957, shoppers could buy Kitty Litter at supermarkets. The product freed cat lovers from having to constantly let their cats in and out (and in and out and in again).

Although cleaning the litter box is no one's favorite chore, the invention of cat litter revolutionized feline living. For the first time, cats could stay inside 24/7. This helped spark a huge shift in what humans thought about how cats should live, as well as how they affected the environment. And with no need to be walked, cats became the preferred lower-maintenance furry pets, perfect for modern living.

Cats Finally Get Their Fill . . .

As cats began to shift away from their primary role as rodent control, they needed an alternate food source. In the 1860s, a British manufacturer named James Spratt began selling the first dog biscuit. The first canned dog food was introduced in 1922, and canned cat food wasn't far behind, coming on the scene in the 1930s. Rationing of metals and meat during World War II popularized the use of kibble and other dried pet foods. In the decades since, as average American prosperity soared and prepared foods for humans became popular, many manufacturers turned to pet food as a way to make use of the meat and fish by-products that we wouldn't eat.

Extrusion machines also revolutionized the production and popularity of dry kibble. Purina developed a process that cooked together wet ingredients, pushed the product through a machine, and baked the resulting niblets. More convenient, cheaper, and less potentially wasteful than canned food, dry kibble became a popular option for both cats and dogs.

. . . and They're Healthier Than Ever

For cats, however, dry food had two major drawbacks: It had a higher percentage of carbohydrates than wildcats had evolved to eat, and the low moisture content could create kidney issues if cats didn't drink enough water. And over the years, as humans began to grow more conscious of the ingredients and preparation of their own food, they turned that same scrutiny to what their pets were eating.

Bolstered by this interest, scientists are increasingly improving food for cats and other pets. In the past decade, the market for prescription brands certified by veterinarians, as well as "boutique" pet food, has exploded and is now worth an estimated 30 percent of the $40 billion annual U.S. market. Many of these premium brands are marketed as organic and natural. One commercial cat food is even designed to inhibit cat allergens and help manage allergic reactions in humans to Fel d1, a protein in cat saliva.

All these changes since the 1950s have transformed our view of cats from mobile mouse hunters to pampered pets. Spay/neuter programs and the transition to an indoor lifestyle means that we have curtailed kitty's independence. Keeping our cats healthy, then, will require us to bring the benefits of outdoor living to our indoor pets.

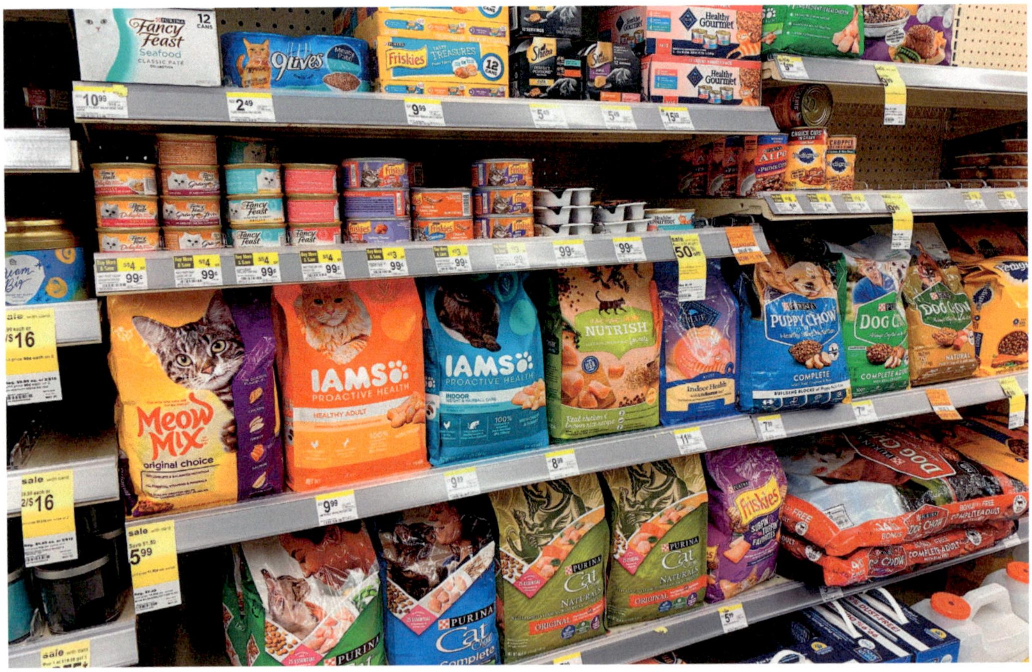

A typical aisle in a grocery or pet supply store (left) features cat food in a wide variety of flavors and even for specific diets and developmental stages.

A selection of cooked protein and vitamins for kittens (opposite) showcases 1950s-style marketing of cat food targeting their doting humans.

Canned Food
Vitamins
Powdered Milk
Liver
Lamb
Wheat Germ
Salad Oil
Eggs
Ground Beef
Hors
Kidney
A
B

CATS IN NATURE

About 80 million cats currently live in American households, and 30 percent of these pet kitties spend some of their time outside. Estimates on how many cats in the United States are free-roaming or feral vary widely, with some sources citing numbers in the tens of millions. Some cat rescues believe that the United States has as many unhomed or "community" cats as it does felines with a fixed address. Many cats—even those with humans to call their own—still spend much of their time roaming at will. With such large numbers of cats going outdoors, even relatively infrequent hunting can have a dramatic impact on local wildlife.

HUNTER CATS HAVE A MAJOR IMPACT ON THE WORLD'S ECOSYSTEMS.

A 2020 study found that both free-living and pet cats kill an estimated 1.3 to 4 billion birds each year, and 6.3 to 22.3 billion small mammals. And it's not just birds and rodents. Scientists from the University of Georgia found that feral cats frequently hunted small reptiles and amphibians, as well as large insects. And because cats don't always bring their prey home, their humans may not realize how much their kitties hunt. Those worldwide tallies show that hunter cats raise serious ecological issues, which humans who love them and care about wildlife and the natural world need to address.

Small wonder that some ecologists have placed cats at the top of the list of the world's most harmful invasive species. Part of the solution is as simple as not allowing cats to roam outdoors. But managing large free-roaming cat populations has proved far more controversial.

Hunting when outdoors comes naturally to cats, but they can happily adapt to indoor living with some help from us.

LOVING THE GREAT INDOORS

STUDIES SHOW THAT indoor cats live longer and are healthier than outdoor cats. But no one wants to be cooped up forever—so what's a cat lover to do?

Cat behavior experts recommend bringing as much of the great outdoors inside as possible. Place cat perches near windows to allow kitty to watch the world go by. Strategically placed bird feeders and insect- and wildlife-attracting shrubs offer up-close entertainment to keep cat minds sharp. Adequate playtime with toy prey can let cats unleash their inner predator without endangering wildlife.

Many cat lovers take their cats outside on a leash and harness. Although some cats love this, be cautious. Skittish kitties may be overwhelmed, and cats can slip out of even snug-fitting harnesses. Before heading out with your cat, talk with your vet to ensure kitty has all necessary vaccines and preventive medicines and is microchipped, collared, and tagged.

BENEFITS OF LIVING WITH CATS

Why would anyone spend so much time and money on family members who don't share our DNA and seem all too happy to find fault with their accommodations? Some people may argue that Fido performs a variety of functions, from guarding livestock and pulling sleds to assisting in hunts and acting as a service animal for humans in need. Although cats may help to keep rodents away, that's not their main purpose in many households anymore. For cat lovers, the softness of our pets' fur and the rumble of their purrs more than make up for the hair balls and predawn demands for food.

Improving Our Health and Wellness

Scientists have documented plenty of other benefits, too. Although many studies document the general benefits of living with pets, fewer have focused specifically on cats. One Australian study found that people who live with cats were happier, less stressed, and better able to face life's problems than their feline-free counterparts. Another study found that people who lived with cats had fewer negative emotions like sadness and jealousy, and that singles with cats had fewer bad moods than those with a cat and a partner. The cat people also appear to be more trusting and socially sensitive.

Cats also help us cope with stress. When researchers tested how well 120 married couples with cats dealt with a series of stressful tasks, they found that people performed best when their cats were with them. They also had lower blood pressure and recovered faster physically when their cats were nearby. Even watching short clips of cat videos online can decrease anxiety and depression and promote happiness and contentment (though doing so to procrastinate is less beneficial).

PEOPLE LIVING WITH CATS ARE HAPPIER, LESS STRESSED, AND COPE BETTER WITH LIFE'S PROBLEMS.

Living with cats can also have physical benefits. Scientists followed 4,435 people over 13 years, and they found people who had cats were less likely to die from a heart attack during the study period. Listening to a purring cat doesn't just reduce stress—it also lowers blood pressure and can help

One of life's great pleasures is sharing the love and trust of a warm and sleepy cat cuddling up in her human's arms.

bones, muscles, and joints heal faster after injury. These benefits can crop up quickly, too. When researchers studied people who had just adopted a cat, they found their number of minor health complaints, such as headaches, backaches, and colds, dropped after just one month, although this waned over time.

And though cats themselves can be a source of allergies, numerous studies show that children who grow up with pets, especially when very young, are less likely to develop allergic reactions later.

FAMOUS CATS IN MODERN HISTORY

Despite their humble beginnings, a select few cats have made history and become cultural icons.

For many of us, childhood brought tales and nursery rhymes about the three little kittens who lost their mittens and the cat and the fiddle (along with the cow who jumped over the moon). If we didn't live with cats, these fictional felines were often a young person's first exposure to cats. Although no one has yet met the old lady who swallowed a fly and then swallowed a dog to catch a cat, these long-standing stories tell us just how tightly cats are interwoven into human society.

Children's books like *Alice's Adventures in Wonderland*, with the mysteriously disappearing Cheshire Cat, and Dr. Seuss's mischievous tale *The Cat in the Hat* brought unforgettable cats to new generations. The popularity of these books—and their cat characters—created new icons for both cat lovers and kids at heart.

Introducing Felix and Félicette

In the television age, animated cats became as popular as their literary counterparts. One of the first to appear was Felix the Cat, a black-and-white animated kitty introduced in "Feline Follies"

A vintage picture from a book of nursery rhymes is a reminder that cats have long been part of the stories we share with children.

HOW THE *MANEKI-NEKO* CONQUERED THE WORLD

In Japan, only one cat can rival Hello Kitty's popularity: the *maneki-neko*, or beckoning cat. This figurine is traditionally a calico Japanese bobtail with a raised paw, but she can be in various colors and different styles. She is often called the "lucky cat." The maneki-neko's raised paw is said to bring good luck to its humans, and her popularity and mystique have spread worldwide. The beckoning gesture of the maneki-neko with paw and claws pointing down and back mimics a welcoming wave in Japanese culture. For non-Japanese markets, her paw faces and waves upward, a gesture more familiar to Westerners.

Historians still argue about when the maneki-neko first appeared—and why. Many of these stories center around an impoverished shop owner who took in a hungry stray cat despite the human's financial predicament. The cat repaid the kindness by beckoning customers into the shop, which brought prosperity to the owner.

High diddle diddle,
The cat and the fiddle,
The cow jumped over the moon;
The little dog laughed
To see such craft,
And the dish ran away with the spoon.

NOT SURPRISINGLY, ADVERTISING BEGAN TAPPING INTO THE FELINE ZEITGEIST.

in 1919. Felix first starred in short, silent films that made him a household name (and brought moviegoers to theaters). Later, he appeared in comic books and, by the late 1920s, in speaking roles.

This tuxedo cat also inspired the name of the first cat in space, a female black-and-white cat named Félicette. As part of the French space program, Félicette became the sixth mammal to reach outer space on October 18, 1963. Two months of training accustomed her to the darkness and close confines of the space capsule, and to the necessary handling by scientists. Félicette was preceded by the U.S.S.R.'s space dog, Laika, who didn't survive her trip to space, and the American chimps, Ham and Enos, who did. Implanted electrodes recorded Félicette's brain activity during the launch, and the kitty landed safely after being launched by a French rocket.

From Tom and Jerry to Town Hall

Beginning in 1940, the animated series "Tom and Jerry" joined Felix at cinemas. The series featured the comic cat-and-mouse games between Tom and his rodent rival, Jerry. Despite each other's often violent efforts, complete with mousetraps and falling sledgehammers, neither cat nor mouse ever manages to get the best of the other. The successful format followed the duo onto television and home theater releases.

Garfield the lasagna-loving cat (and nemesis of dog Odie) began life as a comic strip in 1978. The popularity of the sarcastic kitty led to a cartoon TV show in the 1980s followed by a full-length movie, with sequels release in 2006 and 2022. In 1974, Hello Kitty launched in Japan, creating a global phenomenon that maintains its popularity even now.

Morris, the world's most finicky feline, was introduced to an unsuspecting world in 1968. Successor orange cats still help sell 9Lives®.

Advertisers began tapping into the feline zeitgeist. Not surprisingly, kitty mascots headlined advertising campaigns for newly launched lines of commercially prepared cat food. The orange kitty Morris remains on cans of 9Lives® cat food after more than 50 years. And anyone who heard the commercial for Meow Mix® kibble in the 1970s and 1980s can probably still sing along and quote the tagline ("Cats ask for it by name!").

A buff orange cat named Stubbs was elected mayor of Talkeetna, Alaska, in 1997, where he served the people for 20 years until his death in 2017. Although few cats have held formal office (their lack of thumbs and fondness for catnip making them unelectable), cities and towns around the world have united in support and love of a local cat. And in today's divisive world, this underscores just how popular cats remain.

DECODING THE CAT-INDUSTRIAL COMPLEX

Until very recently, ancient Egypt could be considered to be the human society to have reached "peak cat." Egyptians in antiquity had cat deities and cat-inspired rituals. And although the loss of a beloved pet can bring on an extended period of mourning, even the most vocal cat lovers don't generally mummify their deceased pets (although you can opt to have your kitty taxidermied as a permanent keepsake). But the ancient Egyptians never imagined the internet.

In 2005, Steve Chen, co-founder of YouTube, uploaded the first cat video—entitled "Pajamas and Nick Drake"—a 30-second clip of his cat playing with a rope; the first viral cat video appeared a year later. The website I Can Has Cheezburger launched in 2007, allowing users to create memes by adding their own captions on top of funny cat pictures. What resulted was a long-standing sensation that spawned its own language ("LOLcat") and the ongoing tradition of user-generated memes that can be shared on social media.

These early websites laid the groundwork for the later popularity of any number of cats that have dominated the internet. Maru, a Scottish Fold cat living in Japan, was one of YouTube's

Grumpy Cat®'s face launched a million memes. She had feline dwarfism and an underbite.

Fans of Lil BUB raised money to sequence her genome. Geneticists found she had a rare mutation also found in one mouse and 15 humans.

first celebrities, and the first celebrity cat in general. Grumpy Cat®, aka Tardar Sauce, became famous because a genetic mutation gave her a perpetually annoyed expression. Grumpy Cat® was joined by the likes of Lil BUB and Henri, le Chat Noir.

What a Cat Is: Ours

More recently, rights to the ownership of the animated GIF of Nyan Cat was auctioned off as a non-fungible token, or NFT, in 2021 for $587,000.

The popularity of cats is so central to the internet that when a team of engineers at Google X developed an artificial intelligence algorithm that showed 10 million YouTube images to 16,000 computer processors, the first thing the algorithm did was teach itself to recognize a cat. The scientists never told the computers, "This is what a cat is." They didn't even try to teach a computer to learn what one looked like. Instead, the internet had so many cat images, the computer figured it out.

The question is, why?

One theory is that computer programmers are just more likely to be cat people, which is reflected in the content they make. But our obsession with cats predates computers and will probably last long after the internet has evolved into something completely different. So cats conquered the internet the same way they took over the world: mainly because we love them.

CAT 2.0

In just 10,000 years, cats have come in from the wild and accompanied humans on their journeys around the globe. Now they have ascended to rule the digital world. These recent exploits are but an evolutionary eyeblink in comparison to the feline's 11 million years of evolution in a world free of humans.

Even during the thousands of years cats have spent in the company of people, we have had little influence on when and with whom they have mated. Physically, domestic cats vary relatively little from their wild relatives. Unlike dogs, who have been bred for specific jobs over hundreds or even thousands of years, most cat breeds are less than 200 years old. Although dogs can range from tiny teacup Chihuahuas to enormous Great Danes, most cats are fairly uniform in size and shape. All of this helps to underscore just how recently cat domestication took place.

Breeds Apart

Only recently have breeds like Ragdolls and Birmans appeared, as well as the curly-haired Cornish Rex and the hairless sphinx. Other breeds like Persians are older (their origins can be traced back to the Middle East in the 1600s), but they're still relative youngsters compared to many dog breeds. And though some dog breeds have marked differences in instincts and personality, most cat breeds differ primarily in appearance, not temperament or size. Cat breeders are still actively creating new varieties, including domestic/wildcat hybrids. The Bengal was originally an Asian leopard cat crossed with a house cat. The Savannah is a cross between an African serval and a domestic cat. Some think that both breeds are controversial because of their wild ancestry.

How Will Cats Adapt?

Now, thanks to the invention of cat litter, a growing understanding of the environmental damage wrought by free-roaming cats, and an increase in spay/neuter programs, humans are placing unprecedented pressures on their feline friends. Humans are now asking cats to live indoors 24/7, away from large numbers of their own species, and without the constant stressors of reproduction or hunting for food. Most cat experts are confident that cats will adapt—after all, they've done so before. The question is: What kind of cat will they become?

One possibility is that the cats of the future will be larger than today's felines. As the Vikings plundered

These young Savannahs are a hybrid cat breed introduced in 1986 (above). They are now the largest domestic cats.

The hairless sphinx was developed through selective breeding over the last 60 years (left).

Europe more than 1,000 years ago, they used cats both as pest control and as a source of fur. When a group of Scandinavian archaeologists compared the size of cat pelts from the time of the Vikings to the bodies of modern-day kitties, they found that 21st-century cats were 16 percent larger.

What made these results so unusual is that domestication usually shrinks an animal. Dogs are a quarter smaller than wolves; livestock animals, such as cattle, sheep, and goats, are also smaller than their wild counterparts. And it's not just the availability of energy-dense kibble that has led to bigger cats. Scientists have also documented larger body sizes in free-roaming and feral cats in the Australian bush. Although these cats might steal human food scraps, they aren't all being regularly fed by people.

Of course, cats won't magically grow to be Godzilla-size any more than they will suddenly evolve opposable thumbs (so Fluffy will have to continue to rely on her teeth to tear into that bag of treats while she waits for the Thumb Fairy to appear). But continued contact with humans may also help cats grow friendlier.

Our Lions in the Living Room

By nature, cats tend to be more solitary and standoffish than dogs, which have a long evolutionary history as pack animals rather than lone hunters. Ongoing contact with humans may have helped give friendlier cats a survival advantage, making them more likely to pass on more sociable genes to their offspring. Over time, this might have helped create a more relaxed pet—without ruining one of the most endearing parts of living with a cat: that when your kitty cuddles, you know it's for real.

CATS CAN LIVE SIDE BY SIDE WHEN THE CIRCUMSTANCES ARE RIGHT.

When scientists compared the social behaviors and personalities of domestic cats with clouded leopards, snow leopards, African lions, and Scottish wildcats, they found that house cats weren't exactly suited for group living. They tended to be domineering, neurotic, and impulsive—not exactly the traits you want in a friend.

However, the personalities of house cats were most like those of the more sociable African lion, which has figured out a way to live in groups. This suggests that cats have the raw material to adapt to a more sociable future. And though large cat colonies may not be the same as lion prides, lions do show that cats can live side by side when the circumstances are right.

As humans spend more time learning about cat psychology and behavior, this effort creates an opportunity to improve our relationship. By understanding what makes cats tick—what motivates them, what makes them happy, what they can and can't tolerate—we can ensure that our cats have the best lives possible. Whether it's indoor enrichment or outdoor adventures on a harness and leash, we've only scratched the surface of the human-cat relationship. Whatever happens in the future, our cats will remain by our sides, beloved companions to our fullest possible existence.

A solitary snow leopard is spotted by remote camera amid flurries in Hemis National Park, Ladakh, Jammu and Kashmir, India.

THE SECRET LIFE OF CATS

Carrie Arnold

PRODUCED BY
NATIONAL GEOGRAPHIC PARTNERS, LLC
1145 17th Street NW
Washington, DC 20036-4688 USA

ISSN 2160-7141

Published by Meredith Operations Corporation
225 Liberty Street • New York, NY 10281

Printed in the USA

Special thanks to Steph Selice, Jill Foley, Jackie Brown, Carolyn A. Shea, and Marshall Kiker.

ILLUSTRATIONS CREDITS

Cover, Andrew Marttila; 0-1, Andrew Marttila; 2-3, Joel Sartore/National Geographic Photo Ark; 4-5, Sam Abell/National Geographic Image Collection; 6-7, Charlie Hamilton James/National Geographic Image Collection; 8, VPC Photo/Alamy Stock Photo; 9, Mauricio Anton/Science Source; 10, Joel Sartore/National Geographic Image Collection; 11 (UP), Pete Oxford/National Geographic Image Collection; 11 (LO), Minden Pictures; 12, griepeies/Shutterstock; 13, Joe Petersburger/National Geographic Image Collection; 14, Marser/Getty Images; 15, Beverly Joubert/National Geographic Image Collection; 16, Ermolaev Alexander/Shutterstock; 17, Melville B. Grosvenor/National Geographic Image Collection; 18-19, Jodi Cobb/National Geographic Image Collection; 19, Ed George/National Geographic Image Collection; 20, daphnusia/Shutterstock; 21, Silas Manhood/Alamy Stock Photo; 22, adogslifephoto/iStock/Getty Images; 22-23, Andrew Marttila; 24, Tetiana Kovalenko/Alamy Stock Photo; 25, Mitsuaki Iwago/Minden Pictures/National Geographic Image Collection; 26, sdominick/Getty Images; 27, Leslie Underhill/Elsa Bassiere/National Geographic Image Collection; 28, Samuel Whitton/Alamy Stock Photo; 29, Jean-Michel Labat/Science Source; 30, EEI_Tony/iStock/Getty Images; 31, Beverly Joubert/National Geographic Image Collection; 32-33, Geralyn Shukwit/National Geographic Image Collection; 33, Jeremy Woodhouse/Getty Images; 34, Andrew Marttila; 35, Cyndi Monaghan/Getty Images; 36, Nils Jacobi/Shutterstock; 37, Melissa Farlow/National Geographic Image Collection; 38-39, Bill Luster/National Geographic Image Collection; 40, Magui RF/Shutterstock; 41, Raul Touzon/National Geographic Image Collection; 42-43, Walter Chandoha/National Geographic Image Collection; 44, Vivienstock/Shutterstock; 45, Annie Griffiths/National Geographic Image Collection; 46, John Eastcott and Yva Momatiuk/National Geographic Image Collection; 47, Walter Chandoha/National Geographic Image Collection; 49, Akimasa Harada/Getty Images; 50, Life on white/Alamy Stock Photo; 50-51, Lynn Johnson/National Geographic Image Collection; 52, Eric Isselee/Shutterstock; 53, imageBROKER/Alamy Stock Photo; 54, nevodka/iStock/Getty Images; 54-55, Tierfotoagentur/Alamy Stock Photo; 56-57, James L. Stanfield/National Geographic Image Collection; 58, cynoclub/iStock/Getty Images; 59, Tadamasa Taniguchi/Getty Images; 60, gerard lacz/Alamy Stock Photo; 61, GlobalP/iStock/Getty Images; 62, Melanie DeFazio/Stocksy; 63, Sissie Brimberg/National Geographic Image Collection; 64, Juniors Bildarchiv GmbH/Alamy Stock Photo; 64-65, Tierfotoagentur/Alamy Stock Photo; 66, Gail Heaton/Alamy Stock Photo; 67 (UP), Joel Sartore/National Geographic Image Collection; 67 (LO), Andrew Marttila; 69, John Greim/LightRocket/Getty Images; 70-71, akg-images/John Hios; 72, Richard Barnes/National Geographic Image Collection; 73, National Maritime Museum, Greenwich, London; 74-75, Stringer/picture alliance/Getty Images; 75, Indianapolis Museum of Art/Gift of J. W. Alsdorf /Bridgeman Images; 76, Willard Culver/National Geographic Image Collection; 77, Kharbine-Tapabor//Shutterstock; 78, Paul Nicklen/National Geographic Image Collection; 79, Edward Lowe Foundation; 80, Joni Hanebutt/Alamy Stock Photo; 81, Three Lions/Stringer/Getty Images; 82, Luke Adams/Alamy Stock Photo; 83 (UP), Paul Adams/Alamy Stock Photo; 83 (LO), justinecottonphotography/iStock/Getty Images/Getty Images; 85, Chalit Saphaphak/Stocksy; 86, Jon Helgason/Alamy Stock Photo; 87, Niday Picture Library/Alamy Stock Photo; 88-89, Bettmann/Contributor/Getty Images; 90, Charles Sykes/Bravo/NBCU Photo Bank/NBCUniversal/Getty Images; 91, Chelsea Guglielmino/Getty Images; 92, Kucher Serhii/Shutterstock; 92-93, Wirestock, Inc./Alamy Stock Photo; 94-95, Steve Winter/National Geographic Image Collection; 97, Seth Casteel/National Geographic Image Collection.

No matter how well we think we know our cats and how much we love them, they still can and will surprise us. Often.

Made in the USA
Las Vegas, NV
11 November 2026

57876998R00059